Advances in Intelligent Systems and Computing

Volume 962

The series "Advances in Intelligent Systems and Computing" contains publications on theory, applications, and design methods of Intelligent Systems and Intelligent Computing. Virtually all disciplines such as engineering, natural sciences, computer and information science, ICT, economics, business, e-commerce, environment, healthcare, life science are covered. The list of topics spans all the areas of modern intelligent systems and computing such as: computational intelligence, soft computing including neural networks, fuzzy systems, evolutionary computing and the fusion of these paradigms, social intelligence, ambient intelligence, computational neuroscience, artificial life, virtual worlds and society, cognitive science and systems, Perception and Vision, DNA and immune based systems, self-organizing and adaptive systems, e-Learning and teaching, human-centered and human-centric computing, recommender systems, intelligent control, robotics and mechatronics including human-machine teaming, knowledge-based paradigms, learning paradigms, machine ethics, intelligent data analysis, knowledge management, intelligent agents, intelligent decision making and support, intelligent network security, trust management, interactive entertainment, Web intelligence and multimedia.

The publications within "Advances in Intelligent Systems and Computing" are primarily proceedings of important conferences, symposia and congresses. They cover significant recent developments in the field, both of a foundational and applicable character. An important characteristic feature of the series is the short publication time and world-wide distribution. This permits a rapid and broad dissemination of research results.

** Indexing: The books of this series are submitted to ISI Proceedings, EI-Compendex, DBLP, SCOPUS, Google Scholar and Springerlink **

More information about this series at http://www.springer.com/series/11156

Jessie Chen
Editor

Advances in Human Factors in Robots and Unmanned Systems

Proceedings of the AHFE 2019 International Conference on Human Factors in Robots and Unmanned Systems, July 24–28, 2019, Washington D.C., USA

Springer

Editor
Jessie Chen
U.S. Army Research Laboratory
Orlando, FL, USA

ISSN 2194-5357   ISSN 2194-5365   (electronic)
Advances in Intelligent Systems and Computing
ISBN 978-3-030-20466-2   ISBN 978-3-030-20467-9   (eBook)
https://doi.org/10.1007/978-3-030-20467-9

This Springer imprint is published by the registered company Springer Nature Switzerland AG
The registered company address is: Gewerbestrasse 11, 6330 Cham, Switzerland

Advances in Human Factors and Ergonomics 2019

AHFE 2019 Series Editors

Tareq Ahram, Florida, USA
Waldemar Karwowski, Florida, USA

10th International Conference on Applied Human Factors and Ergonomics and the Affiliated Conferences

Proceedings of the AHFE 2019 International Conference on Human Factors in Robots and Unmanned Systems, held on July 24–28, 2019, in Washington D.C., USA

Advances in Affective and Pleasurable Design	Shuichi Fukuda
Advances in Neuroergonomics and Cognitive Engineering	Hasan Ayaz
Advances in Design for Inclusion	Giuseppe Di Bucchianico
Advances in Ergonomics in Design	Francisco Rebelo and Marcelo M. Soares
Advances in Human Error, Reliability, Resilience, and Performance	Ronald L. Boring
Advances in Human Factors and Ergonomics in Healthcare and Medical Devices	Nancy J. Lightner and Jay Kalra
Advances in Human Factors and Simulation	Daniel N. Cassenti
Advances in Human Factors and Systems Interaction	Isabel L. Nunes
Advances in Human Factors in Cybersecurity	Tareq Ahram and Waldemar Karwowski
Advances in Human Factors, Business Management and Leadership	Jussi Ilari Kantola and Salman Nazir
Advances in Human Factors in Robots and Unmanned Systems	Jessie Chen
Advances in Human Factors in Training, Education, and Learning Sciences	Waldemar Karwowski, Tareq Ahram and Salman Nazir
Advances in Human Factors of Transportation	Neville Stanton

(continued)

(continued)

Advances in Artificial Intelligence, Software and Systems Engineering	Tareq Ahram
Advances in Human Factors in Architecture, Sustainable Urban Planning and Infrastructure	Jerzy Charytonowicz and Christianne Falcão
Advances in Physical Ergonomics and Human Factors	Ravindra S. Goonetilleke and Waldemar Karwowski
Advances in Interdisciplinary Practice in Industrial Design	Cliff Sungsoo Shin
Advances in Safety Management and Human Factors	Pedro M. Arezes
Advances in Social and Occupational Ergonomics	Richard H. M. Goossens and Atsuo Murata
Advances in Manufacturing, Production Management and Process Control	Waldemar Karwowski, Stefan Trzcielinski and Beata Mrugalska
Advances in Usability and User Experience	Tareq Ahram and Christianne Falcão
Advances in Human Factors in Wearable Technologies and Game Design	Tareq Ahram
Advances in Human Factors in Communication of Design	Amic G. Ho
Advances in Additive Manufacturing, Modeling Systems and 3D Prototyping	Massimo Di Nicolantonio, Emilio Rossi and Thomas Alexander

Preface

Researchers are conducting cutting-edge investigations in the area of unmanned systems to inform and improve how humans interact with robotic platforms. Many of the efforts are focused on refining the underlying algorithms that define system operation and on revolutionizing the design of human–system interfaces. The multifaceted goals of this research are to improve ease of use, learnability, suitability, and human–system performance, which in turn will reduce the number of personnel hours and dedicated resources necessary to train, operate, and maintain the systems. As our dependence on unmanned systems grows along with the desire to reduce the manpower needed to operate them across both the military and commercial sectors, it becomes increasingly critical that system designs are safe, efficient, and effective. Optimizing human–robot interaction and reducing cognitive workload at the user interface require research emphasis to understand what information the operator requires, when they require it, and in what form it should be presented so they can intervene and take control of unmanned platforms when it is required. With a reduction in manpower, each individual's role in system operation becomes even more important to the overall success of the mission or task at hand. Researchers are developing theories as well as prototype user interfaces to understand how best to support human–system interaction in complex operational environments. Because humans tend to be the most flexible and integral part of unmanned systems, the human factors and unmanned systems' focus considers the role of the human early in the design and development process in order to facilitate the design of effective human–system interaction and teaming.

This book will prove useful to a variety of professionals, researchers, and students in the broad field of robotics and unmanned systems who are interested in the design of multisensory user interfaces (auditory, visual, and haptic), user-centered design, and task-function allocation when using artificial intelligence/automation to offset cognitive workload for the human operator. We hope this book is informative, but even more so that it is thought-provoking. We hope it provides inspiration, leading the reader to formulate new, innovative research questions, applications, and potential solutions for creating effective human–system interaction and teaming with robots and unmanned systems. Five sections are presented in this book:

Section 1   Advanced Methods for Human-Agent Interfaces;
Section 2   HRI Applications in the Workplace and Exoskeletons;
Section 3   Human-Robot Communications and Teaming;
Section 4   Human, Artificial Intelligence, and Robot Teaming;
Section 5   Human Interaction with Small Unmanned Aerial Systems.

Each section contains research papers that have been reviewed by members of the International Editorial Board. Our sincere thanks and appreciation to the Board members as listed below:

July 2019   Jessie Chen

Contents

Human-Robot Communications and Teaming

Human, Artificial Intelligence, and Robot Teaming

Human Interaction with Small Unmanned Aerial Systems

Advanced Methods for Human-Agent Interfaces

Vision Ship Information Overlay and Navigation "VISION" System

Jessica Reichers[✉], Nathan Brannon, Joshua Rubini, Naomi Hillis, Katia Estabridis, and Gary Hewer

Naval Air Warfare Center – Weapons Division,
1 Administrative Circle, China Lake, CA 93555, USA
{jessica.reichers,nathan.brannon,joshua.rubini,
naomi.hillis,katia.estabridis,gary.hewer}@navy.mil

Abstract. Modern naval vessels, marvels of systems engineering, combine a myriad of complex solutions into a single, sophisticated machine. Personnel responsible for the safe operation of these ships are required to parse, filter, and process a large array of information in order to make key decisions. Safe navigation requires better processing and interpretation of significant quantities of data, and ensuring this information is made available in an easily consumable fashion.

Augmented reality provides a means to solve this problem, using customizable interfaces and overlays of information on the world to help make intelligent decisions in dangerous, congested situations. The team's Optimal Trajectory and other path planning algorithms can refine these navigational aids, accounting for changes in weather, current, and tides.

Our team successfully demonstrated an augmented reality solution, giving additional situational awareness during transit to sea and into port.

Keywords: Augmented reality · Decision making · Human factors

1 Introduction

Futuristic human machine interfaces envisioned in movies like Iron Man, now seem achievable with emergence of mixed reality headsets. With an Augmented Reality (AR) Head Mounted Display (HMD) developers can create the illusion of seeing inside bodies or seeing air flowing in a room [1, 2]. One such device, the Microsoft Hololens, digitizes the world with tangible hologram interactions.

In May 2018, the team was challenged to create an application for improved situational awareness in three months that would work aboard a moving sea vessel. They created SitaViewer, a software and hardware suite, that handled visualization of stationary objects outside of the bridge while the ship sailed through a channel. Aricò et al. demonstrated large scale situational awareness for air traffic control with the Hololens [3], however it was unknown whether the Hololens would work in a room while the vehicle was moving.

There are also known limitations of the device such as the bulkiness and the small field of view (FOV) which the team had to design around [4]. Other groups speculated

J. Chen (Ed.): AHFE 2019, AISC 962, pp. 3–14, 2020.
https://doi.org/10.1007/978-3-030-20467-9_1

the device would evolve, and recently Microsoft released a patent for a new Near-Eye-Display (NED) which suggests that the current version's projection system will be removed in the next hardware iteration of the device [5]. Additional improvements are expected to be a result of a recent contract the company has with the Army [6]. These modifications and improvements are necessary for the headset to facilitate improved safety and reduced cognitive load for ship operators. Unfortunately, there is a lack of research being done using the Hololens on a moving vehicle.

This paper describes our process to building and validating a prototype called SitaViewer on the ship's bridge. This paper is laid out as follows: the background addresses what kind of information it could display for the ship and why there is a lack of development in this area. The methods section shows the hardware and software systems that the team developed. The results section discusses the tests and demonstrations, with the conclusions and future work section following and describing features that are planned for future iterations.

2   Background

Piloting a ship is an information intensive discipline; the following information needs to be available and modifiable in real-time: local knowledge, transit-specific knowledge and knowledge of ship handling. Grabowski explained the previous information is used for the three tasks of piloting: track keeping, maneuvering, and collision avoidance, as well as adherence to procedures and good practice developed over years of ship handling [7]. Maneuvering through a channel introduces risk because movement is constricted by the channel. The risks include running aground, entering restricted water zones and damaging the ships on the shore [8].

Microsoft Hololens research groups have created proof-of-concept tools for professional use in stationary environments since 2016. Although, no application exists for a room on a large slow-moving vehicle such as a sea vessel, possibly due to the headset's design and standards. Its Health&Safety regulations warn against use while operating motor vehicles because the images can distract the user and occlude their view [9]. The device was designed to work under the assumption that its environment is stationary with the user moving relative to the space. In order to display realistic 3D computer images directly in the world involves maintaining awareness of the physical space and the head of the user relative to that space. This is done by creating a digital world map with the its cameras and sensors and tracking head movement with an inertial navigation system (INS). All of which is filtered by their proprietary Holographic Processing Unity (HPU) chip before being processed on the GPUs [10].

Groups have added extra functionalities to the Hololens, but no external sensors would allow it to differentiate movement of the surroundings from movement of the user. At this time there was no way to change what the Hololens detects as user movement [11]. Granted, it is possible to modify the headset with external hardware for additional capabilities such as computer vision or awareness of occluded areas [12, 13]. The company does allow some sensor access were a user can turn a flat surface into a virtual touch pad [14].

3 Problem

The team was tasked to create and validate a novel situational awareness Hololens display as a standalone prototype for channel navigation in three months. The display needed to realistically overlay real world objects such as buoys while the ship was piloted into or out of port and inform the user of the distance each buoy was from the hull. An initial test was required to establish the viability of the headset in a moving vehicle. Afterwards, the problem consisted of three parts: creating a mock test environment, the user experience (UX) and ship testing.

In this iteration the team did not have access to the ship's systems that provided state information due to accreditation and authorization constraints. They required a separate data source to provide GPS location, speed and heading. This system had to be portable and mountable for temporary installments in a vehicle during test and communicate with the HMD via wireless Ethernet.

This device required graphic and software development to be done within the Unity 3D environment. The Unity 3D scene required a network connection to receive state information, populated the buoys in the world and handled user inputs. The graphics in the scene needed to overlay buoy images over real-world buoys. In addition, the Unity 3D scene in the Hololens performed the necessary calculations to show distances between buoys and the ship's hull.

4 Methods

As discussed previously, the team developed a complete software and hardware augmented reality solution showing channel navigation buoys using the Microsoft Hololens headset. In addition, the team designed software to include markers in between the buoys as virtual "lane lines" showing the channel boundaries.

4.1 Hardware System Design

The overall system design consisted of three major components: a state data source (Odroid minicomputer and autopilot), the base station (laptop running Linux OS), and the headset (Hololens). Other supporting equipment included a custom power supply for the data source, and a network router to supply wireless communication to the headset.

Data Source

The team could not access the ship's internal state data directly, therefore they needed some alternative means to localize the headset in the world. This localization facilitated a synchronized "digital overlay" on the physical world that they populated with objects whose approximate positions were known, such as the channel buoys. In order to accomplish this, the team needed accurate position, heading, and velocity data.

Position data was needed so that the digitally displayed items appeared in the view at their correct locations, within a reasonable margin for error. While the team did not expect the digital representations to exactly match the actual real objects, due to expected error in sensed position and heading measurements, they did not want the error to be so great such that the device became more of a hindrance than an aid to the user.

Heading data was needed to counteract the problem of the headset not detecting that the vehicle was mobile. The team needed a way to tell the device to rotate objects in the view when it had not detected any rotation of the device with respect to its internally mapped environment.

Velocity data was useful as an internal check to changes in position, and as information for the user. It was not specifically needed for any calculation regarding updates to the world, as those were handled simply as updates to current location using position data, while direction was updated using heading data.

The team was already experienced at using small hobby autopilots for controlling quad-rotors, and these devices supplied the necessary information within usable error tolerances. The device used was a Pixhawk autopilot, connected to a pair of GPS receivers. The Pixhawk sent updates to the Odroid containing position, heading, and velocity at approximately 10 Hz, which were then relayed to the base station. Error in the civilian GPS measurement was calculated to be less than five meters, which at the distances we were expecting object to be viewed at would not introduce an unduly jarring discrepancy.

It was noted at the time of selection that the Pixhawk's heading measurement was via a magnetic compass, and that this would most likely not function well in the magnetic interference generated by the ship. This observation was ultimately borne out as discussed in the next sections.

The state data was read over a serial connection by a program running on the Odroid computer, which was chosen due to its portability, low power requirements, and ability to connect both a serial cable and Ethernet cable. The program on the Odroid established the connection to the base station and Pixhawk, then simply repackaged the information coming from the Pixhawk and transmitted them over Ethernet to the base station. Figure 1 shows the data source physical connection diagram.

Base Station

The system design needed a means to communicate state and other pertinent information to the headset from a data source. The Hololens was designed to communicate with other electronics wirelessly, which required wireless communication to the headset within the ship, essentially a Faraday cage, from the data source. The unreliability of receiving a wireless signal from outside the bridge required the wireless broadcast to be within the bridge as well, so some sort of intermediary between the GPS receiver/Odroid and the headset was necessary.

Using a laptop computer as the intermediary gave the team significant flexibility as to what computations should be performed by the headset, and what could be exported to a more powerful, external computer. The computer selected as the base station contained: 32 GB RAM, 3.7 GHz Intel I7, and an nVidia GeForce GTX 1080 Mobile GPU, running Ubuntu Linux v. 18.04.

In addition, the base station had a full development suite, including the Unity 3D development environment, allowing the team to minimize the amount of equipment needed at the test and demonstration sites.

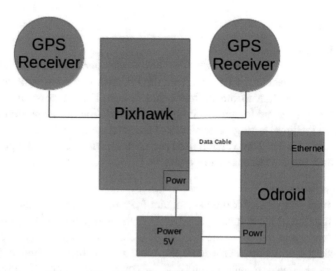

Fig. 1. Data source diagram. This schematic shows the data source used to generate position and heading messages for use by the Hololens headset.

The base station used in the test described in this paper used custom software to receive the state information from the data source via wired ethernet connection, and transmit it to the headset over wireless. The program allowed multiple headsets to connect to a single data source.

Headset

When requested to explore augmented reality for shipboard situational awareness and navigation assistance, the Hololens was the natural choice due to its onboard computational capability, ease of interface, and spatial mapping and interactivity. The Hololens is truly unique in this regard, in that it allows the user to effectively turn their entire surroundings into a digital canvas with its ability to anchor windows and 3D objects to surfaces that it detects, not simply display information in a limited viewing window.

Another built-in feature of the Hololens is the ability to record what it sees along with any digital addition. This feature is of particular interest to the Navy, following the recent collisions involving US naval vessels. This has sparked an interest in technologies that can provide a visual record of events as they occur, which the Hololens can do, albeit at a significant cost to visual display quality.

Finally, the Hololens was the logical choice for the team to use for this project, as they already had experience developing applications for it, had two of the devices on hand, and the truncated time line would have made developing a similar capability for a different headset very challenging.

4.2   Software Design

Odroid Software

The solution required a program on the Odroid that performed two tasks. First, it packaged and broadcast telemetry data from the Pixhawk to the base station. Second, it monitored the connection to the Pixhawk and gracefully shut down the data stream when it detected it was no longer receiving data. A variation of this program ran on the base station to facilitate testing other pieces of the solution without having to connect to the external data source. That version used recorded position data from a previous local test to simulate the movement of the platform.

Base Station Software

The base station software took the data stream from the data source, and rebroadcast it over wireless to the headset. It accepted data from the data source and stored it, which it then read from to rebroadcast to the headset. If there was more than one message since the last read, the software rebroadcast the most recent, then cleared old data. The software was designed to allow multiple outgoing connections to allow multiple headset users at once. While it was essentially just repeating messages sent to it from the data source, the base station also managed the incoming connection requests from headsets, which allowed it to send different data to different headsets based upon user preferences.

Headset Software

By far the most challenging piece of the software puzzle was the software running on the headset. This program took position, speed, and heading data in a stream from the base station, and constructed a virtual world overlay showing buoy positions, channel boundaries, and other state information.

The headset virtual overlay was built as a scene in the Unity 3D development environment. The team chose Unity 3D due to already having an API for the Hololens, and their experience using it for other projects. In addition, Unity 3D has a rich set of resources and built in libraries for three-dimensional graphic rendering, as well as an advanced object handling infrastructure that was leveraged in the project.

There were some specific challenges introduced by the problem the team solved; namely, the camera position and orientation representing the Hololens in the Unity 3D scene was not controlled by the developer. The Hololens mapped its environment using onboard sensors and computation, and then sent position updates to Unity 3D based upon its perceived change of orientation and/or position within its surroundings. Under normal use-case circumstances, this would be a welcome reduction of work that the developer had to do and would correctly move and orient the camera in the world. This solution assumes, however, that the "world" is stationary. For the actual problem, the "world" was the deck of a ship and was moving with respect to the larger environment. It was this larger environment that the team were interested in augmenting, so the Hololens not recognizing the relative movement was a serious issue.

The team solved this by translating the position messages into moving the world around the camera. This allowed the compound movement of the Hololens moving within its mapped surroundings, and the surroundings moving within the larger ocean, to be correctly rendered on the display. The objects are positioned and then translated around the camera, effectively showing the correct position of the objects in the world.

The Hololens software was preloaded with approximate positions of buoys which were then referenced by the Unity 3D scene. In order to show a particular object, the Unity 3D software needed to either receive the object's position from the base station, or have its position read either from a file, or as part of the program code. The team encountered significant problems trying to setup a data file on the device, and as such decided to "hard code" the positions of objects as part of a script in the Unity 3D scene.

Due to the large number of buoys marking the channel, and the limited field of view (FOV) that the Hololens provided, the team needed a means to show buoys that were not within that FOV. The team used a mechanism inspired by other simulated reality programs, video games, and heads-up displays, that showed small arrow-like indicators along the border of the view to show that if the user turned that way, there would be an object of interest in that direction.

An additional feature was added which showed a dotted line representing the interpolated lines between the buoys as an analog to lane lines on highways. This idea was accepted and developed, with the result shown in Fig. 2.

Of significant concern was the smoothness of the translation and rotation when the user turned to look in a different direction. The team did not want the software trying to "catch-up" to the user's new orientation, causing significant jumps of objects in the view, nor did they want the error in the GPS measurement to cause the objects to move erratically. Significant effort was made to ensure smooth, even translation and rotation.

Heading

The Pixhawk autopilot used a magnetic compass to report heading data. This compass was rendered inoperable in proximity with the magnetic field generated by the ship, which caused significant error in the positions of objects in the view.

The solution the team used was a restricted Kalman filter, which sampled prior position data to compute an average heading based upon current and past positions. The filter tracked three moving windows of position data, and randomly sampled the

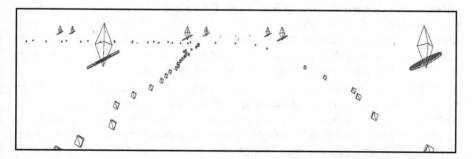

Fig. 2. View of overlay in Unity 3D. This shows the screen elements that would be displayed on the Hololens as an overlay of the real world. The small squares are the virtual "lane lines" marking the interpolated channel boundaries between buoys. The diamonds with disks beneath them represent buoys.

first and third windows to compute heading from positions in those windows. Under the assumption that the positions were reported at a uniform rate, it also computed the speed of the ship, although this was redundant as the Pixhawk's speed computation was deemed to be accurate regardless of the compass interference. This amounted to a very accurate heading-over-ground measurement (accuracy was within $0.1°$ during straight-line navigation with $\pm$ five meters of error in the GPS positions, and within $0.5°$ once a turn was detected and the time between sampled positions reduced).

4.3   Testing and Validation

The proposed demonstration was overlaying channel buoys while piloting into and out of port. Specifically, to show the buoys in the channel leading to Puget Sound Naval Shipyard from Puget Sound. Additionally, if necessary, the team could show buoys marking the channel in another harbor.

Local Test

The team used a small personally owned vehicle, and surveyed locations near the development site, as means to test the software package prior to the demonstration. The test site used telephone poles as physical objects to represent on the overlay, both due to the regular distance between them, and having something substantial that the team could actually see easily to compare the accuracy of the overlay to. Additionally, the device was walked down the test site to provide a longer time-window, fine-grained test of the software.

As expected, the software struggled to recognize the turning of the vehicle that it was operating within, and this helped refine the solution to that particular problem. Subsequent car tests showed that the headset would eventually register the turn, which caused problems with how the platform was rendered in the view. The team eventually decided to simply not show the platform until this problem could be addressed in future iterations.

Transit Demonstration

The team demonstrated the initial version of the product in the agreed upon use-case on a transit from Bremerton, WA, to San Diego, CA, while piloting out of Bremerton to Puget Sound, and while piloting into San Diego harbor. During both of these demonstrations, they spent significant time setting up the external data source, the communications hardware necessary to get the data from the data source to the Hololens and initializing the Hololens for use with the system.

The crew of the ship was very helpful in providing the team with the necessary materials and guidance for all of the equipment to work smoothly, and there were few problems encountered along the way. One of these problems was that they had to run an Ethernet cable through a watertight closure, and during a drill, the closure was secured, which severed the cable. This problem was foreseen by the team, however, and they were well equipped with a spare cable.

Fig. 3. Overlay shown prior to entering port in San Diego. The black circles are drawn around small grey triangles that appear to show where buoys are located, in this case, on the other side of Point Loma. Note the ship's speed indicated in the small grey box below the phone, and the distance to the closest buoy in the view in the grey box on the left side, above the window. These boxes effectively delimit the viewing area available to show the overlay.

5   Results

Results of internal tests and simulations showed that the team could take a GPS position and heading stream and produce a virtual overlay on the Hololens device that both tracked with actual objects in the world and provided useful information in a heads-up display format. The car and walking tests outside the development arena did show some initial problems with the solution, both in setup time, sensitivity to synchronization issues, and sensitivity to the error in GPS measurement.

Solutions to these issues were proposed, designed, and implemented prior to the scheduled product demonstration and were deemed sufficient to go ahead with the demonstration. The team flew to Bremerton, WA, boarded the ship, and spent five days on board. The initial demonstration of the transit out of Bremerton had several issues, of which the failure of the Pixhawk heading was the most significant. Still, they were able to show the lane lines and certain buoy locations, despite the difficulties with the data source. This demonstration was not as successful as it could have been, but it did spark a discussion with the crew on what additional features would be most useful, after which they agreed to add a GUI element that displayed the speed-over-ground of the ship.

The team spent the next few days refining the display, implementing an algorithm to derive heading from position history, adding the requested feature, and preparing for the entry into San Diego harbor. The new heading algorithm was tested extensively against simulated position data and determined to be accurate enough to be usable.

During the transit, two pieces of equipment failed, the Ethernet cable connecting the data source to the base station, and the wireless router which managed the wireless network connecting the base station and the Hololens. The team replaced the Ethernet cable with a backup and worked a solution to the router using a spare laptop as network router.

Piloting into port in San Diego went much more smoothly, requiring less setup time due to no need to calibrate the compass on the Pixhawk, and with other wrinkles discovered in the Bremerton transit solved as well. The team demonstrated the Hololens with the overlay correctly showing buoy locations, even those obscured by the deck and/or terrain features such as Point Loma, as seen in Fig. 3. Several crew members had suggestions for improvements, which were documented. Overall, the augmented reality solution with the Hololens was well received, and the team was invited back to the ship later in 2019 to show the next iteration incorporating the crew's requests.

6 Conclusions and Future Work

The results of the demonstration in San Diego confirmed that augmented reality as an aid to situational awareness is a viable tool on board a naval vessel. The crew had several specific requests for functionality, including:

1. Showing the closest point of approach of a contact based upon its course and speed with respect to the ship's heading and speed.
2. Showing the track down centerline as opposed to the channel lane lines. This would be more generally applicable, and give specific information for things like where to maneuver to stay on the planned course, etc.
3. Show the distance and time to the next planned maneuver.
4. Show the current heading and speed.
5. Show the set and drift (current and tide) which is affecting the movement of the ship, along with depth beneath the keel.
6. Give an audible warning and highlight a dangerously close object with a visible aura if/when such a situation arises.
7. Show a virtual chart. This object could be anchored to any of the Hololens' mesh surfaces that it detects, basically allowing the user to always have the current chart available.
8. Record events as they occur.

The solution worked well on the bridge while the ship was moving. The issue persisted regarding camera position and orientation, and a solution or modification to these in Unity 3D and/or on the Hololens itself would greatly ease development of more applications where the headset is enclosed on a moving vehicle. One physical issue encountered during both testing and demonstration was that the holographic display was significantly muted in very bright lighting, and the team solved this by placing a semi-opaque sunscreen on the visor of the headset. A more robust, well-designed sunshade developed specifically for the Hololens would be well received.

The team plans to investigate incorporating other functionality into the system, including integrating algorithms for optimal trajectory planning to suggest maneuvers that account for the dynamics of the ship, surface traffic, and conditions such as set and drift [15]. These computations can be done on the base station, which would then pass it along with the other state information updates. Dynamic models of the ships have been developed with sufficient fidelity for use in the optimal trajectory planning software [16].

Machine learning techniques combined with the real-time video from the Hololens and computer vision algorithms may be used to identify surface contacts that the person wearing the Hololens sees. This could be beneficial in situations where the ship is highly constrained in its ability to maneuver, giving early warning of a possible attack during these vulnerable situations.

Other uses include providing information during specific scenarios such as damage control, flight operations, or underway replenishment. During these, the Hololens' ability to act as both heads-up display, digital canvas, and x-ray vision could provide critical information to leadership personnel at an unprecedented speed.

There are still many challenges associated with overlaying an accurate digital representation on the real world, particularly when the observer is constantly moving. As the technology for augmented reality continues to improve, the team will continue to explore solutions which improve the fidelity of the virtual overlay, ease-of-use of the device, and overall experience.

References

1. Kim, M., Yi, S., Jung, D., Park, S., Seo, D.: Augmented-reality visualization of aerodynamics simulation in sustainable cloud computing. Sustain. **10**, 1362 (2018)
2. Wu, M., Chien, J., Wu, C., Lee, J.: An augmented reality system using improved-iterative closest point algorithm for on-patient medical image visualization. Sens., 1–13 (2018)
3. Aricò, P., et al.: Human-machine interaction assessment by neurophysiological measures: a study on professional air traffic controllers. In: EBMC 2018, 40th International Engineering in Medicine and Biology Conference, Honolulu (2018)
4. Baumeister, J., et al.: Cognitive cost of using augmented reality displays. IEEE Trans. on Vis. Comp. Graph. **23**, 2378–2388 (2017)
5. Robbins, et al.: Compact optical system with mems scanners for image generation and object tracking. US Patent 10,175,489 B1, 8 Jan 2019
6. Bloomberg LP. https://www.bloomberg.com/news/articles/2018-11-28/microsoft-wins-480-million-army-battlefield-contract
7. Grabowski, M.: Research on wearable, immersive augmented reality (WIAR) adoption in maritime navigation. J. Navig. **68**, 453–464 (2015)
8. VEAT. https://fortress.wa.gov/ecy/publications/documents/1508012.pdf
9. Microsoft. https://www.microsoft.com/en-us/hololens/legal/health-and-safety-information
10. Popular Mechanics. https://www.popularmechanics.com/technology/gadgets/a15324/how-microsofts-hololens-works/
11. Microsoft.   https://docs.microsoft.com/en-us/windows/mixed-reality/device-portal-api-reference

12. Garon, M., Boulet, P., Doiron, J., Beaulieu, L., Lalonde, J.: Real-time high resolution 3D data on the hololens. In: IEEE ISMAR, 1–3 (2016)
13. Qian, L., Plopski, A., Navab, N., Kazanzides, P.: Restoring the awareness in the occluded visual field for optical see-through head-mounted displays. IEEE Trans. Vis. Comp. Graph. **24**, 2936–2946 (2018)
14. Xiao, R., Schwarz, J., Throm, N., Wilson, A., Benko, H.: MRTouch: adding touch input to head-mounted mixed reality. IEEE Trans. Vis. Comp. Graph. **24**, 1653–1660 (2018)
15. Robinson, D.R., Mar, R.T., Estabridis, K., Hewer, G.: An efficient algorithm for optimal trajectory generations for heterogeneous multi-agent systems in non-convex environments. IEEE Robot Auto. **3**, 1215–1222 (2018)
16. Skjetne, R., Smogeli, O., Fossen, T.: A nonlinear ship manoeuvering model: identification and adaptive control with experiments for a ship model. Mod. Ident. Contro. **25**, 3–27 (2004)

Task Engagement Inference Within Distributed Multiparty Human-Machine Teaming via Topic Modeling

Nia Peters[(✉)]

Battlespace Acoustic Branch, 711th Human Performance Wing,
Air Force Research Laboratory, Dayton, OH, USA
nia.peters.1@us.af.mil

Abstract. Research in *Intelligent Awareness Systems (IAS)* focuses on designing systems that are aware of their current environment by monitoring human interactions and making inferences on when to engage with human counterparts. A potential gap is task engagement inference for distributed human-machine teaming. The objective of this paper is a proposed intelligent awareness system via task topic modeling for task engagement inference within these domains. If the system has information on "what" teammates are discussing or the task topic, it is better informed prior to engaging. The proposed task topic model is applied to two simulated multiparty, distributed teaming interactions and evaluated on its ability to infer the current task topic. For both tasks, the model performs well over the random baseline, however the performance is degraded for interactions with more robust dialogue. This work has the potential of informing the development of intelligent awareness systems within distributed multiparty teaming and collaborative endeavors.

Keywords: Task engagement inference · Human-systems integration · Human-machine teaming · Topic modeling

1 Introduction

Human-machine teaming aims to meld human cognitive strengths and the unique capabilities of smart machines to create intelligent teams adaptive to rapidly changing circumstances. Human-machine teaming has attracted the attention of the U.S. military and commercial sector for domains such as human-robot teams, unmanned aerial vehicle (UAV) operations, operator control stations, supporting overworked and multitasking hospital personnel, air traffic control stations, commercial and military pilots in cockpits, and human-computer technical support teams.

One major problem in human-machine teaming is a lack of communication skills on the part of the machine such as the inability to know *when* to communicate information to human teammates. Research approaches in Intelligent Awareness Systems (IAS) offer a potential solution to this problem. IAS focuses on designing systems that are aware of their current environment by monitoring human interactions and making inferences on when to engage with humans. In an attempt to make these engagement decisions, studies in the process of *task engagement* between humans and

J. Chen (Ed.): AHFE 2019, AISC 962, pp. 15–24, 2020.
https://doi.org/10.1007/978-3-030-20467-9_2

machines have been explored within the context of multiparty closed-world (Goffman 1963; Kendon and Ferber 1973) and more recently in open-world (Bohus and Horvitz 2009) *situated* interactions where the surrounding environment can provide rich contextual information for managing and systematizing the interaction.

A potential gap within the IAS literature is *task engagement inference* within distributed human teaming endeavors such as air traffic controller/pilot or unmanned aerial vehicle (UAV) operator/ground troop teams where environmental cues in situated interactions are not available. An approach to narrow this gap is a proposed intelligent awareness system that can make task engagement inferences using only spoken language, a primary and convenient modality within distributed teaming interactions. Task engagement is a collaborative construct and teammates are working on collaborative tasks. From a spoken language perspective, knowing who is speaking, who they are addressing, and what they are speaking about allow the system to make more informed decisions about their engagement within distributed multiparty teaming. Speaker and addressee identification in multiparty distributed interactions, can be addressed using call-signs or channel designations and is therefore out of the scope of this paper.

The primary focus of this research is on what teammates are speaking about or the current *task topic*. The intuition behind this approach is that if the system has information regarding the task topic, it is better informed prior to engaging with information relevant to the ongoing topic, a request to switch topics, or a signal to continue the task for off-topic discussions. The overall objective of this paper is to illustrate a proposed *task topic model* within multiparty distributed human-machine teaming interactions as a proposed task engagement intelligent awareness system. The problem of task topic modeling is formalized into a topic identification or topic modeling problem. If the intelligent awareness system can monitor the speech over a radio channel shared by two or more humans and infer the current task or context, it could potentially be better informed of the current topic of discussion and provide assistive or supporting information.

This work has the potential of informing the development of an intelligent awareness system for distributed multiparty teaming and collaborative endeavors where the system is aware of the current topic of discussion and can provide information relevant to the current task topic or detect "off topic" conversation and better infer when to engage with human collaborators. In expanding on approaches within the IAS literature, there is the potential to give a machine the ability to monitor the communication and interaction of human teammates, and make inferences on when to engage bringing us closer to automating communication within human-machine teaming.

2   Background

Research in *Intelligent Awareness Systems (IAS)* focuses on designing systems that are aware of their current environment by monitoring human interactions and making inferences on when to engage with human counterparts (Horvitz and Apacible) (Fogarty, Hudson, and Lai, Examining the robustness of sensor-based statistical models of human interruptibility) (Fogarty et al. 2005). Recent literature suggests that capturing *task engagement* is important for creating reliable models on when to engage with humans (Hudson et al. 2002; Perlow 1999; Seshadri and Shapira 2001). Studies in the process of engagement between humans and machines have been explored within the context of multiparty closed-world exchanges using hand gestures (Kendon and Ferber 1973), gaze and mutual attention (Argyle and Cook 1976), spatial trajectory and proximity (Hall 1966) (Kendon), non-verbal cues (Goffman 1963; Kendon and Ferber 1973), as well as open-world situated interactions (Bohus and Horvitz 2009) where the surrounding environment can provide rich context beneficial in managing and organizing the interaction. In the system proposed by (Horvitz and Apacible), speech and facial recognition cues are leveraged to determine which users are currently engaged in an ongoing multi-player game and the system makes inferences on whether to engage with users with information related to the current task. The situated nature of these tasks allows the physical environment to provide useful contextual information. Systems have the opportunity within these interactions to leverage collective information about physical features and relationships such as gestures, facial expressions and the absence or presence human participants. "Physicality and embodiment can provide salient affordances that can be used by a system to support the engagement process" (Bohus and Horvitz 2009).

A potential gap within the current literature is task engagement inference within distributed human teaming endeavors where the physical information may not be available and offer the system salient information to assist in making engagement decisions. Since task engagement has been proposed for situated environments, the aim of this research is to propose an intelligent awareness system for human-machine teaming distributed interactions. Leveraging spoken language such as speech over a radio channel to model task topics is proposed as an initial approach in addressing the limitations within the IAS literature in inferring task engagement within multiparty distributed teaming endeavors.

3 Data Collection

To simulate a distributed multiparty teaming scenario, two experimental designs are presented: the Tangram Task (TT) (Peters et al. 2019, pp. 783–788) and the Uncertainty Map Task (UMT) (Peters et al. 2019, pp. 783–788). The objective of the TT and UMT data collection is to simulate a dual-user information alignment task to represent a distributed teaming collaboration where teammates speak over a radio communication channel to align their knowledge from two different perspectives and accomplish a common goal. If the goal is to integrate a machine teammate into this interaction, the system would listen to the human interaction and make decisions on when to engage with humans with information related to the ongoing exchange in a way that is assistive and least disruptive to the overall exchange.

3.1 Tangram Task

The Tangram Task (TT) involves two teammates corresponding over a push-to-talk (PTT) communication network to arrange abstract shapes (Tangrams) within a column into corresponding order to simulate aligning knowledge from two different perspectives. Tangrams are objects composed of geometric shapes so some objects can be described based on their shape composition or what the object resembles. Figure 1 illustrates the Graphical User Interface (GUI) for the TT from each teammate in a dual-user team.

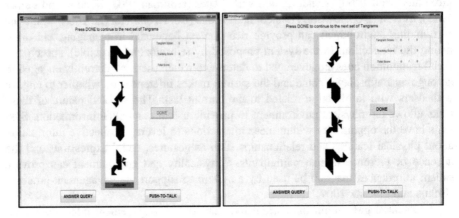

Fig. 1. Dual-user Tangram Task (TT) GUI

The TT is proposed as a knowledge alignment task that allows teammates to establish a back and forth dialogue of uncertainty to successfully arrange their Tangrams into corresponding order. A succession of Tangram columns with random Tangram shapes is presented to both teammates. The pictures in each column are generated from Tangrams shapes that are similar in appearance or classes that are used for task topic labels. The task topic labels are ABSTRACT, BOATS, PEOPLE, ANIMALS, and BIRDS. There are 250 Tangram images (50 images × 5 classes).

The primary conversational exchange between teammates is the teammates taking turns describing their shapes. The speaker describes their shapes, the partner arranges the shapes accordingly and responds with a confirmation statement such as "got it" or follow-up questions prior to confirming their understanding. A new set of Tangrams is generated after both teammates press the DONE button and this process continues until 15 min has elapsed.

3.2   Uncertainty Map Task

The Uncertainty Map Task (UMT) is another simulated multiparty, distributed teaming exchange like the Tangram Task where the aim is for teammates to align their knowledge by describing a target house from two perspectives and agreeing on the target house. In the UMT, two humans communicate over a push-to-talk network to describe their respective interfaces, ground their knowledge, and identify a target house. Figure 2 illustrates the dual-user UMT GUI.

In the UMT there are 4 aerial maps and 12 houses for each aerial map (12 × 4 = 48 target houses). Each target house has 4 different perspectives. The UMT

Fig. 2. Dual-user Uncertainty Map Task (UMT) GUI

is comprised of 4 different tasks that are integrated into the design of the experiment to establish task topic labels. The task topic labels are Aerial Target and Street View Identification (AERIAL_TARGET_SV_ID), Street View Target and Aerial Identification (SV_TARGET_AERIAL_ID), Street View Target and Street View Identification (SV_TARGET_SV_ID), and House Label (HOUSE_LABEL). Figure 2 is an example of the Street View Target and Street View Identification (SV_TARGET_S-V_ID) where one teammate has the target house from one street view perspective and the other teammate must identify the same target house from another street view perspective.

Like the Tangram task, the teammates work on a series of randomly generated UMT interfaces. Usually the teammate with the "target" perspective describes the house form their perspective and the teammate with the "ID" view asks follow-up questions or confirms their understanding prior to pressing the DONE button. The teammate with the "target" and "id" perspectives is also randomly generated throughout the experiment. The teammates go through a series of randomly generated UMT tasks until 15 min has elapsed.

4 Features and Modeling

The problem of task topic modeling is formalized into a topic identification or topic modeling problem. In this approach, the raw audio of the push-to-talk (PTT) utterances from the Tangram Task (TT) and Uncertainty Map Task (UMT) are processed by a speech-to-text unit to extract the words from the audio for each utterance. The words in each utterance are mapped into feature space associated with the likelihood that the word sequence is associated with a topic using the Latent Dirichlet Allocation (LDA) topic modeling approach and classified into its respective class using the maximum entropy classifier. Mallet (McCallum 2002) was used to build and evaluate this Lexical-Maximum-Entropy (MAXENT) model. This approach outperformed other experimental approaches and is hence presented here. A 10-fold cross validation method is used to generalize the model with 80% training and 20% testing distribution for each fold. For both the TT and UMT the data was down sampled to ensure equal distribution across task topic labels.

The performance metrics used to determine how well the model performs are recall (sensitivity), precision, and the F1 score. *Precision* is defined as the number of true positives divided by the number true positives plus the number of false positives. *Recall* is defined as the number of true positives divided by the number of true positives plus false negatives. Recall expresses the ability to find all relevant instances in a dataset and precision expresses the proportion of the data points the model says were relevant are indeed relevant. The F1 score is the harmonic mean of the recall and precision.

This approach in using task topic modeling assumes there are certain words associated with certain topics. In the Tangram Task, one could imagine there are words specific to each of the Tangram groups. For example, there may be words more prevalently used to describe the BIRD group such as *wing* or *beak* that are distinct from words used to describe PEOPLE such as *hands* or *hair*. Conversely there is the potential of word overlap between these two topics such *head* and *legs*. This is an illustration of a potential limitation of a topic modeling approach.

5   Results and Discussion

To evaluate the performance of the proposed task topic model, a visualization of the feature vectors (via keywords) extracted using topic modeling and the performance of the MAXENT classifier using metrics such as precision, recall, and the F1 score are presented. Figure 3 illustrates the top 20 keywords associated with each topic cluster for the TT. The words highlighted in red are words that are in every topic cluster.

These are words that probably carry less topical information such as articles, prepositions, fillers, etc. This overlap of words between topics is reflected in the features and makes utterances in different task topics less distinct effecting the overall classification performance. Simply removing these words could aid in tuning these models.

The words highlighted in green are words that may carry topical information but are present across different tasks. Since Tangrams are geometric shapes, teammates may be using shapes to describe the objects such as the word *triangle*. Additionally, some of the shapes are rotated and for some users these look like completely different objects, but for others they are perceived as previously viewed images that have been rotated. Hence words such as *right, left, upside*, and *down* are common across topic clusters. Without domain knowledge, it is potentially more challenging to identify words that are prevalent within the teaming interaction, but carry less information in discriminating topics a priori. As a result, blindly removing these words as previous mentioned for commonly used words with less information may not be as effective especially in dynamic interactions where the dialogue is more robust. Strategies to address this challenge would afford an opportunity to ignore prevalent, but non-topical words so the model could concentrate on words with topical information that aid in discriminating task topics.

TOPIC I: the, and, guy, then, person, have, you, down, triangle, like, that, got, with, running, upside, this, he's, lady, holding, his

TOPIC II: got, the, bird, and, flying, have, down, upside, then, okay, flamingo, monster, duck, turkey, ness, loch, you, with, seal, crow

TOPIC III: the, and, one, like, then, have, with, first, dog, last, looks, down, it's, second, third, right, left, that, has, triangle

TOPIC IV: the, and, then, have, sailboat, got, with, boat, house, triangle, tree, okay, sideways, upside, apple, like, arrow, down, two, bridge

TOPIC V: the, tree, palm, corner, left, bottom, triangle, and, then, are, slanted, leaves, top, boat, speed, going, okay, it's, have, second

Fig. 3. Topic modeling keywords for the Tangram Task

Finally highlighted in blue are the words that carry topical information for discriminating utterances into different task topics. For example, illustrated in Fig. 3 Topic I could be associated with the PEOPLE class with words such as *guy, person, running, lady,* and *holding*. Topic II could be associated with the BIRDS class with words such as *bird, flying, flamingo, duck, turkey,* and *crow*. Topic IV and Topic V could be associated with either the BOAT or ABSTRACT classes with words such as *sailboat, boat, bridge, tree, apple, leaves.*

Figure 4 illustrates the top 20 keywords associated with each topic cluster for the UMT. For the UMT, the task topic clusters were not as salient as those illustrated in the TT.

TOPIC I : got, have, okay, you, three, pictures, one, think, view, i'm, aerial, what, alright, i've, this, that, houses, sets, just, yeah,

TOPIC II : the, house, and, like, roof, has, it's, that, front, side, there's, driveway, corner, with, this, right, there, one, sidewalk, then

TOPIC III : the, sidewalk, does, that, front, okay, one, out, house, path, kind, into, walk, crosses, wishbone, changes, watch, whereas, forming, continue

TOPIC IV : like, that, see, the, but, you, it's, don't, yeah, know, looks, can, think. one, i'm, and, not, look, really, just

Fig. 4. Topic modeling keywords for the Uncertainty Map Task

Topic I could be associated with the Aerial Target and Street View Identification task with phrases like "three pictures" associated with the identification of a street view house (shown on the left of Fig. 2) where the user was presented with *three* different houses in which one was the target house and commonly used the phase "I have three houses." Additionally, the word "aerial" could align with the teammate describing the aerial view in this same interaction. Also, illustrated across all topics are words such as *house*, *sidewalk*, and *front* which one could imagine could be used across the different task topics. Topic IV seems to be words with little topical information. Future analysis would have to be done to better understand topic modeling within the context of this data collection. For brevity and focus on an approach to task engagement inference within multiparty, distributed teaming, topic modeling is used as tool to map lexical information (words) into a feature space that representative topics to infer task topics.

Table 1 illustrates the multiclass (topics) classification results for the task topic model.

Table 1. Task topic model classification results

Task	Precision	Recall	F1	Baseline	N
Tangram Task	82.7%	79.7%	**81.2%**	20%	12,316
UMT	55.1%	50.5%	55.7%	25%	8,658

The random baseline for each task topic model is 20% for the TT and the 25% for the UMT. From the topic model keyword output, the separation in the words over the topics in the Tangram Task and the overlap of words across the task topics in the UMT task are reflected in the classification performance illustrated in Table 1. The task topic model performs with an F1 score of 81.2% for the TT in contrast to the model for the UMT that performs with an F1 score of 55.7%. One observation is that the Tangram Task is a larger dataset than the UMT, but experiments were performed to randomly down sample the TT dataset to ensure the superior performance was not attributed to larger training data. The results showed no significant difference in the reported accuracy.

A salient point to illustrate is the ability of the model in both tasks to outperform the random baseline. The task topic model for the TT performs **300%** over its random baseline. The model for the UMT performs **116%** over its random baseline. To our knowledge this is a novel approach on a unique dataset so it is safe to compare our modeling results to a random baseline for illustrating system performance. A potential contribution to the literature is that for multiparty distributed human-machine teaming interactions, the following results could serve as a baseline for similar tasks and modeling endeavors.

A few limitations of the study are if the objective is to leverage spoken language from a distributed multiparty interaction to inform a system, it would be useful to better understand not only topic modeling so that we could turn these models to accomplish better performance, but also explore other natural language processing features and modeling strategies. Another limitation is a more comprehensive understanding of these models to better inform and inspire additional approaches. Overall the use of topic modeling to infer task engagement has potential even in interactions with robust dialogue such as UMT and seems to do well in dialogues that are more predictable such as the Tangram task.

6   Conclusion and Future Work

In conclusion, this work aims to contribute to the IAS literature with a proposed intelligent awareness system for distributed, multiparty teaming interactions. A task topic modeling strategy to infer task engagement is proposed since similar modeling strategies are used within the literature in multiparty, situated teaming interactions (Bohus and Horvitz 2009). In distributed teaming interactions, visual and physical attributes are limited, but spoken language is available therefore the proposed system leverages this information to infer task engagement. Here task engagement is defined from the perspective of knowing the current task topic. The formalized approach is a task topic modeling strategy which is evaluated within two multiparty, distributed teaming interactions. For the first teaming task with distinct task topics, the proposed algorithm detects task topics with an accuracy of 81.2% (random baseline 20%) where many of the detection errors are a result of spoken utterances that do not carry much topical information (i.e. fillers). A proposed solution to this issue is to remove words with less topical information prior to building topic models for this task. Within the second teaming task with less distinct task topics, the proposed approach detects task topics with an accuracy of 55.7% accuracy (random baseline of 25%) where majority of the detection errors are caused by the similarities in the spoken language used within similar topics.

The topic modeling strategy proposed in this work is inspired primarily by the intuition that topic modeling approaches could be leveraged to predict task topics as an indicator of when to infer task engagement. In future work, it would be worth using similar modeling methods proposed within the situated multiparty teaming work (Bohus and Horvitz 2009) to inform this work even if the input modality is different. Additionally, in future work the aim is to tune topic models, explore additional spoken and natural language features, and other modeling strategies for informing a system when to engage with human counterparts within distributed teaming interactions.

References

Adams, J.: Human Machine Teaming Research. Vanderbilt University (n.d.). Retrieved from www.vanderbilt.edu

Argyle, M., Cook, M.: Gaze and Mutual Gaze. Camridge University Press, New York (1976)

Bohus, D., Horvitz, E.: Dialog in the open world: platform and applications. In: Proceedings of the 2009 International Conference on Multimodal Interfaces. ACM (2009)

Fogarty, J., Hudson, S., Lai, J.: Examining the robustness of sensor-based statistical models of human interruptibility. In: Proceedings of the SIGCHI Conference on Human Factors in Computing Systems, pp. 207–214. ACM (n.d.)

Fogarty, J., Ko, A., Aung, H., Tang, K., Hudson, S.: Examining task engagement in sensor-based statistical models of human interruptibility. In: Proceedings of the SIGCHI Conference on Human Factors in Computing Systems, pp. 331–340. ACM (2005)

Goffman, E.: Behaviour in Public Places: Notes on the Social Order of Gatherings. The Free Press, New York (1963)

Hall, E.T.: The Hidden Dimension: Man's Use of Space in Public and Private. The Bodley Head, London (1966)

Horvitz, E., Apacible, J.: Learning and reasoning about interruption. In: Proceedings of the 5th International Conference on Multimodal Interfaces, pp. 20–27. ACM (n.d.)

Hudson, J., Christensen, J., Kellogg, W., Erikson, T.: "I'd be overwhelmed, but it's just one more thing to do": Availability and Interruption in Research Management. In: Proceedings of the ACM Conference on Human Factors in Computing Systems, pp. 97–104 (2002)

Kendon, A.: Spatial organization in social encounters: the F-formation system. Conducting Interaction: Patterns of Behavior in Focused En- counters, Studies in International Sociolinguistics. Cambridge University Press (n.d.)

Kendon, A., Ferber, A.: A description of some human greetings. Comp. Ecol. Behav. Primates **591**, 12 (1973)

McCallum, A.K.: Mallet: a machine learning for language toolkit (2002). Retrieved from http://mallet.cs.umass.edu

Perlow, L.: The time famine: toward a sociology of work time. Adm. Sci. Q. **44** 57–81 (1999)

Peters, N., Bradley, G., Marshal-Bradley, T.: Task boundary inference via topic modeling to predict interruption timings for human-machine teaming. In: International Conference on Intelligent Human Systems Integration, pp. 783–788. Springer, Cham, San Diego (2019)

Seshadri, S., Shapira, Z.: Managerial allocation of time and effort: the effects of interruptions. Manage. Sci. **47**, 647–662 (2001)

Decision Making Using Automated Estimates in the Classification of Novel Stimuli

Amber Hoenig[✉] and Joseph DW Stephens

North Carolina Agricultural and Technical State University,
1601 East Market St., 27411 Greensboro, NC, USA
ashoenig@aggies.ncat.edu, jdstephe@ncat.edu

Abstract. In large-scale, remotely operated systems of autonomous vehicles, human operators' situation awareness will depend on their use of multiple information sources which may include target classification estimates provided by the system. This experiment assessed to what degree participants relied on a likelihood estimate to assist in the classification of novel stimuli in varying levels of uncertainty. Participants were trained to classify two sets of novel visual stimuli, then classified variations of the stimuli with the aid of an estimate displaying the likelihood of belonging to either group. The results showed that participants were able to integrate the automated estimate into their classification responses, and as the level of uncertainty increased, the average reliance on the automated estimate also increased. The findings show that training participants to identify new stimuli, then presenting participants with a likelihood estimate in conjunction with the visual stimuli may facilitate situation awareness in conditions of uncertainty.

Keywords: Human factors · Human-Systems integration ·
Unmanned vehicles · Decision making · Representation of uncertainty

1 Introduction

Operations of unmanned vehicles provides a wide range of challenges in human factors, decision making, and situation awareness [1, 2]. Humans are an integral part of operations of unmanned vehicles, particularly in decisions which involve safety or cannot ethically be made by a computer [3]. Humans are more resilient than computers in their ability to make decisions with incomplete or uncertain information [3]. In dynamic and uncertain environments, people can adapt to unforeseen circumstances, whereas current technology must be improved to provide better resilience in unfamiliar situations for computer systems [1, 4]. On the other hand, computer systems can be connected to large-scale sensor systems, providing and integrating massive amounts of data [4] which are beyond humans' ability to interpret [5]. These data must be appropriately managed to overcome data overload, allowing the human in the loop to comprehend the relevant information and make appropriate decisions [6–8].

Operators of unmanned vehicles experience great difficulty with maintaining sustained levels of cognitive engagement in supervisory targeting veto tasks, but operator

© Springer Nature Switzerland AG 2020
J. Chen (Ed.): AHFE 2019, AISC 962, pp. 25–35, 2020.
https://doi.org/10.1007/978-3-030-20467-9_3

selection, training, and system design can improve operators' ability to maintain focus [3]. Appropriate interface design features can decrease mental fatigue, lower response times, assist the operator in processing information, and prevent accidents or loss of systems [9]. This study explores a method to assist remote operators of swarms of unmanned vehicles with decision making and to maintain situation awareness.

2 Literature Review

Design of a user interface for swarms of UVs requires careful forethought and planning; each interface is usually tailored to a specific type of mission. However, research regarding the human factors of unmanned vehicle control is still in a very basic stage [1]. User interfaces are designed with functionality in mind but not optimality [1], which is of particular concern in light of the tremendous cognitive resources required to operate a swarm of unmanned vehicles [9]. Interface features which reduce cognitive workload must be identified and implemented to avoid critical performance degradation.

Remote operators of unmanned vehicles typically operate with far fewer sensory cues than those which would be provided in person—typically, the kinetic/vestibular, auditory, and ambient visual cues present in the environment are not provided to remote operators [10]. The human in the loop must instead rely on primarily visual cues to maintain situation awareness and make decisions without the aid of the lost sensory cues. This is cognitively taxing and can lead to mental fatigue and decreased situation awareness or SA. This is defined as "the perception of the elements in the environment within a volume of time and space, the comprehension of their meaning and the projection of their status in the near future" [11]. [7] states that situation awareness "incorporates an operator's understanding of the situation as a whole, which forms the basis for decision making." Loss of this crucial awareness can compromise human operators' ability to successfully complete missions [2].

An additional layer of complexity is created by increasing the number of vehicles a human operator must supervise during a mission. Humans are limited in the number of unmanned vehicles which they can actively control at a given time while still maintaining situation awareness. Authors in [12] found that as the number of unmanned aerial vehicles controlled by experimental participants in a targeting mission increased from one to four, the participants' level of situation awareness decreased significantly. Conversely, as the level of automation of swarms of unmanned vehicles increases, operators can guide or direct the swarm while considering it as a single entity instead of directing the vehicles individually. This reduces workload and decreases the risk of cognitive overload [1]. However, in this case, the data collected by many sensors across the swarm must be fused in order to provide a coherent representation of the environment to the human operator. Such data fusion may be subject to uncertainty, which must be accounted for when modeling the data [5].

Multi-sensor data fusion is concerned with finding a reliable and accurate way to combine multiple inputs from different sensors. It is similar to the concept of human and animal multisensory information integration, which provides valuable information about the environment [13], and thus is important to situation awareness of operators of unmanned vehicles. Information integration is not only assumed but invaluable in operations of swarms of unmanned vehicles; [14] lists data fusion as one of the most crucial tasks of systems with multiple sensors. The combination of data from such a system can provide richer information than could be gathered by a single sensing entity. Mathematical models are used to integrate the sensor data and provide probability estimates for the resulting information.

These data can be used to generate likelihood functions such as how probable it is that a certain event will occur or that a certain object belongs to a given class; thus, an estimate of uncertainty can be provided [5]. Sensor data and a likelihood estimate can be presented to operators of unmanned vehicles to assist in decision making tasks [15].

Maintenance of situation awareness is further complicated by the fallible nature of autonomous vehicles themselves. Unmanned vehicles are subject to failures, communication losses, and sensing imperfections which can create additional conditions of uncertainty [16]. Sensor disturbances in a multi-sensor system can occur due to natural or intentional disturbances [5]. Examples of the former include problems due to atmospheric conditions in optoelectronics and radar errors due to evaporation ducts; intentional disturbances can be created by adversaries wishing to disrupt certain wave forms or wavelengths. Weather, geometrical masking, sensor detection range limits, and many other problems can render some sensors in a system unable to gather accurate data [5, 13]. These difficulties further emphasize the uncertainty of information gathered by swarms of unmanned vehicles. Decision aids can be used to influence operators' decision-making and situation awareness when dealing with uncertain information [15].

A review by [17] found that user interfaces for operations of swarms of unmanned vehicles must be designed with the appropriate level of autonomy in mind; a balance must be attained between reliance on automation and situation awareness. The lowest levels of automation may rely too heavily on the operator to select tasks and direct the swarm, thus creating an unmanageable workload. On the other hand, a level of automation which is too high can decrease situation awareness by leaving the operator in a monitoring position, risking operator boredom and increasing the likelihood that the operator will fail to take action in an emergency situation. A balanced level of autonomy allows the operator to maintain an appropriate level of situation awareness and meet the task goals. The present study mimics a management-by-consent model by presenting points of interest to the participants and asking them to make a decision about the stimuli. Similarly, [1] state that it is crucial to calibrate the level of autonomy of swarms of unmanned vehicles to reduce the number of tasks the operator must complete in a given task. Thus, the operators can maintain situation awareness and limit the decisions they must make to the most important ones, allowing the swarm to take care of most of its own operations. This is in line with a target classification scenario recommended by [5] for crucial decisions in which multiple deployed sensors gather information about the target, identify a set of potential classes which is as small as possible, and offer the potential classes and information about their uncertainty to the human decision maker for classification.

We hypothesize that presenting operators with sensor information from a limited number of unmanned vehicles in conjunction with an estimate gathered from a machine algorithm may allow users to make decisions about autonomous vehicle use without overtaxing their cognitive abilities. Instead of presenting multiple sensor feeds from many autonomous vehicles, we conceptualized training operators to accurately identify novel stimuli, then providing sensor data from only one vehicle as well as a simulated automated machine estimate. This is simulated in the following experiment by the presentation of a stimulus image in conjunction with a numerical or visual decision aid. This model reduces the number of decisions and quantity of stimuli the operator must monitor and simplifies the process, thus increasing the operator's ability to maintain situation awareness.

3 Method

We developed an experiment to measure the performance of participants classifying novel stimuli with the aid of two different types of likelihood estimates. The participants were trained to classify two different groups of unfamiliar objects, then tested on their ability to correctly classify stimuli from those groups aided by an estimate of the stimulus belonging to a given group. The testing phase included stimuli which the participants had been trained to identify as well as new (untrained) stimuli belonging to the same groups. In addition, the stimuli were presented from a frontal view as in the training session as well as from a side view. We expected that the automated estimate would have a significant influence on the participants' decisions, particularly in conditions of uncertainty.

Sixty-two participants from the undergraduate psychology program at North Carolina Agricultural and Technical State University were recruited to participate in this study. Compensation was offered in the form of extra credit for their courses.

The structure of this experiment is based on Lupyan's 2007 experiment evaluating the usefulness of linguistically labeling groups in order to facilitate learning of novel categories [18]. Gauthier's YUFO set are used as the experimental stimuli [19]. The YUFO data set consists of novel three-dimensional objects developed using computer graphics software (see Fig. 1). Each set consists of sixteen highly similar shapes with subtle differences to demarcate them. The results of Lupyan's experiment showed that labels facilitate the learning of new categories, even nonsensical ones.

Fig. 1. Stimuli from the YUFO data set from front and side views.

We chose to use the same sets of stimuli and nonsense labels in order to facilitate learning new categories and to represent unfamiliar types of stimuli that may be encountered in remote unmanned vehicle swarm operations. For example, sensor data from a chemical sniffer or a computer representation of motion may provide readouts which do not readily correlate to human senses [20, 21]. The data set shows unfamiliar objects and encourages participants to rely on their senses and the displayed likelihood estimate to classify the stimuli.

In the training phase, participants were instructed to imagine that they were explorers on a different planet upon which they would encounter different types of aliens. Previous explorers had categorized the aliens as the nonsense words "leebish" and "grecious." Participants were presented with a picture of an alien in the center of the screen and a picture of an astronaut above, below, or beside it; they were prompted to choose whether to move toward or away from the alien. The participants were instructed to use the arrow keys to indicate their response, thus associating a behavioral response with the stimuli.

A total of 144 training trials and 96 testing trials were performed in a full factorial, repeated-measures experimental model. In each trial, participants were presented with a fixation cross for 500 ms and then a 384×384 pixel (2.56×2.56 inches) picture of an alien from the YUFO data set. A 300×400 pixel (1×1.33 inch) picture of an astronaut was placed above, below, or to either side of the alien. Participants responded by indicating whether the astronaut should approach or avoid the alien by pressing the arrow keys to simulate moving toward or away from the alien. Five hundred milliseconds after each response, the same stimulus was displayed in the center of the screen with the name of its group (leebish or grecious) at the top of the picture. Auditory feedback was also provided in the form of a bell for a correct response or a buzzer for an incorrect response. Participants were trained to approach the leebish individuals and avoid the grecious individuals.

During the training session, participants were presented with a set of eight aliens each from the leebish and grecious groups from the YUFO data set [19]. Over the span of 144 trials, participants were trained to identify the aliens as leebish or grecious. Participants indicated their decisions by pressing the "f" key for leebish and the "j" key for grecious. No time limit was imposed for decision making.

At the beginning of the testing portion of the experiment, participants were instructed to identify the aliens with the aid of a machine estimate which was provided to help them make their decision. The estimate was calibrated to be statistically valid–for example, each member of a group comprised of three leebish aliens and one grecious alien would be given an automated estimate of 75% leebish (implying an automated estimate of 25% grecious). In the first iteration of the experiment, the automated estimate was represented as a percentage (e.g. 75% leebish) for the first 30 participants. Based on initial results, we suspected that using a numeric estimate may encourage participants to rely on a rounding rule to make their decisions in the testing phase. This may increase the weight of the percentage estimate by encouraging participants to round 75% up to 100% and 25% down to 0%, making the automated estimate less ambiguous than it actually is. An example of this is one participant's responses; they classified the stimuli as leebish every time the machine estimate displayed "75% leebish" and selected grecious in response to every machine estimate of

50% and 25%. To mitigate these possible issues, we used a simple visual estimate in the second portion. The machine estimate was represented as a circle on a line labeled with the words "leebish" on the left and "grecious" on the right as shown in Fig. 2. The circle was placed at the left quarter, on the center, and on the right quarter of the line to represent likelihoods of 75%, 50%, and 25% leebish and 25%, 50%, and 75% grecious respectively. Figure 2 shows the automated estimate indicating a 75% likelihood of the stimulus belonging to the leebish classification and a 25% likelihood of belonging to the grecious classification.

Fig. 2. The visual automated estimate used in the second portion of the experiment.

A total of 96 trials were run in the testing portion of the experiment. Participants were presented with four aliens which they had been trained to identify as well as four new individuals from each group. To add further uncertainty, the aliens were presented from either a front (trained) or side (untrained) viewing angle. In addition, the experimenters used different stimulus sets in the first and second parts of the experiment (sets one and two for the percentage estimate portion of the research and sets three and four for the visual machine estimate portion) to evaluate whether some of the results were due to particular visual characteristics of the data sets which might be more or less distinguishable and thus alter the results. We expected that as the level of uncertainty increased through providing untrained stimuli and manipulating the viewpoint, the participants would rely increasingly on the machine estimate to guide their decisions.

4   Results and Discussion

The results of the experiment showed that on average, the participants' reliance on the machine estimate increased as the level of uncertainty increased. The condition of least uncertainty in this experiment includes trained stimuli presented from the trained (front) viewpoint. The condition of greatest uncertainty is the untrained stimulus, untrained viewpoint category. This is supported by the mean response accuracies as shown in Table 1 which decrease as the level of uncertainty increases. Figure 3 shows that as the uncertainty of the stimuli increases, participants' responses approach the value of the presented machine estimate. This demonstrates that the participants relied more heavily on the automated estimates as their level of uncertainty increased.

The response proportions were modified using the logit transformation to map the data onto a continuous function. A 2 (category of alien) × 2 (training level of view) × 2 (training level of stimulus) × 3 (machine estimate) repeated-measures ANOVA was performed on the logit transformed response proportions to determine whether any of the independent variables were associated with a change in response patterns.

Table 1. Mean accuracy of responses according to training level of stimulus and viewpoint and mean distance of responses from the value of the automated estimate.

Training level	Mean response accuracy
Trained Stimulus, Trained View	76.23%
Untrained Stimulus, Trained View	70.08%
Trained Stimulus, Untrained View	65.44%
Untrained Stimulus, Untrained View	63.58%

Percent "Leebish" Responses versus Automated Estimates in Varying Conditions of Uncertainty

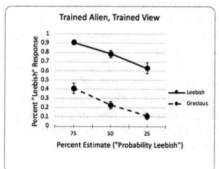

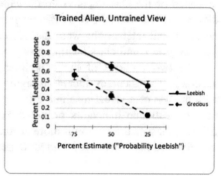

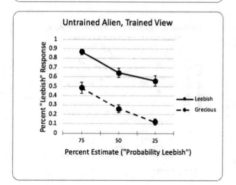

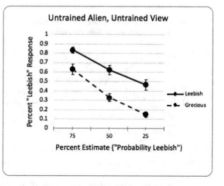

Fig. 3. Percent of stimuli identified as "leebish" versus the percent likelihood indicated by the machine estimate. The four different stimulus conditions represent varying levels of uncertainty. In the most uncertain condition of untrained stimulus/untrained view, the responses are closest to the percentage value given by the automated estimate.

The portion of the study which used a percentage estimate as the machine estimate showed highly significant effects from the machine estimate ($F(2,27) = 34.966$, $p < .001$), stimulus classification (leebish or grecious) ($F(1,28) = 48.478$, $p < .001$), and from the interaction between stimulus training level (trained or untrained alien) and stimulus classification ($F(1,28) = 14.195$, $p = .001$). This part of the experiment also showed significant effects in the stimulus classification × stimulus training level, stimulus classification × viewpoint training level, stimulus classification × machine estimate, and stimulus classification × stimulus training level × machine estimate. This pattern indicates a very strong effect from the stimulus classification and the machine estimate.

The second portion of the experiment which used the visual automated estimate showed results somewhat similar to the first. The machine estimate had a highly significant effect ($F(2,30) = 36.373$, $p < .001$). The stimulus classification had a highly significant effect; ($F(1,31) = 82.368$, $p < .001$). Stimulus classification × viewpoint training level also had a significant effect ($F(1,31) = 10.570$, $p = .003$).

The highly significant effect of the machine estimate indicates that participants were able to integrate the machine estimate into their decision-making process. The significance of the classification group validates that participants were able to learn to differentiate between the two different groups. In the second portion of the experiment, no additional interactions were discovered beyond machine estimate, stimulus classification group, and classification group × training level.

The two portions of the experiment also differed in terms of mean response accuracy. The accuracy increased significantly in the second part of the experiment with the use of a visual machine estimate display (Fig. 4). Although the two types of machine estimate indicators provided exactly the same likelihood information, participants classified the stimuli more accurately (73.75%) when presented with a visual machine estimate than with a numeric machine estimate (63.40%).

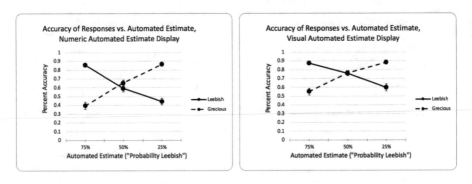

Fig. 4. The mean accuracy of responses for each type of machine estimate.

The difference in responses associated with each type of machine estimate is particularly noticeable in trials in which the machine estimate displayed 50% likelihood for both categories. An automated estimate of 50% in this case essentially offers no information—both classes are equally likely to be the correct choice, so the decision is left to the participant. Given a machine estimate of 50% leebish/50% grecious, the mean accuracy of responses in the visual machine estimate condition reached 76.01%; in the numeric machine estimate category the responses averaged 62.18% (Fig. 4). The latter category's nearness to the given machine estimate may be due to use of a rounding rule (tending to choose one or the other group based on a mental rule of rounding up or down from 50% to find one's choice when uncertain). The visual display may have made it clearer that no information was being offered in this category and thus have encouraged participants to rely on their own their own judgment in these cases. The overall increased accuracy of responses in the visual machine estimate portion of the experiment may be due to a more intuitive representation of the machine estimate.

5 Limitations

Limitations of this work include the number of trials for each participant; although the data were appropriately transformed, the extensive number of combinations of the types of stimuli yield highly discrete data. The participants in the experiment were students drawn from a select field of study; a broader base of participants from different areas of expertise may alter the patterns observed in this research. In addition, although the two sets of stimuli used in the study are similar, the use of one data set with a visual machine estimate and a different data set with a numeric machine estimate could introduce a confounding variable. Future work may include testing the first stimulus sets with the visual automated estimate and the second two stimulus sets with the percentage estimate in order to discern whether there were any effects due to differing characteristics of the data sets. Further testing with a response time limit may provide a more realistic simulation of an urgent unmanned vehicle operation scenario and thus alter reliance on the automated estimate.

6 Conclusions

Presenting participants with an image of a stimulus in conjunction with a simulated automated estimate facilitated decision making in conditions of uncertainty. On average, participants relied more heavily on the machine estimates as uncertainty increased, e.g. given untrained stimuli or stimuli displayed from a side viewpoint. The two types of machine estimates used gave different results; responses to the numeric machine estimate showed decreased accuracy and possible rounding in the condition of equal likelihood of belonging to either class. The visual machine estimate yielded a higher response accuracy and interfered less with accurate classification in the machine estimate condition of 50% leebish/50% grecious.

The results of this study may be applied to design of user interfaces for operations of swarms of unmanned vehicles to facilitate decision making and increase situation awareness. Providing select sensor information about an object of interest in conjunction with an automated classification estimate as simulated in the experiment utilizes the computational power of the system while keeping the human in the loop. This may assist the operator in processing and classifying information, reducing the amount of work the operator must do and thus reducing the likelihood of cognitive overload. By reviewing limited, relevant information about the environment and using an automated estimate, the human in the loop can maintain situation awareness and rely on the estimate as a decision aid in conditions of uncertainty.

Acknowledgments. The authors would like to acknowledge the support from Air Force Research Laboratory and OSD for sponsoring this research under agreement number FA8750-15-2-0116. The U.S. Government is authorized to reproduce and distribute reprints for Governmental purposes notwithstanding any copyright notation thereon. The views and conclusions contained herein are those of the authors and should not be interpreted as necessarily representing the official policies or endorsements, either expressed or implied, of Air Force Research Laboratory, OSD, or the U.S. Government.

References

1. Hocraffer, A., Nam, C.S.: A meta-analysis of human-system interfaces in unmanned aerial vehicle (UAV) swarm management. Appl. Ergon. **58**, 66–80 (2017)
2. Adams, J.A.: Unmanned vehicle situation awareness: A path forward. In: Human Systems Integration Symposium, pp. 31–89 (2007)
3. Taylor, R.M.: Human Automation Integration for Supervisory Control of UAVs (2006)
4. Council, N.R., others: Intelligent Human-machine Collaboration: Summary of a Workshop. National Academies Press, Washington, DC (2012)
5. Appriou, A.: Multisensor data fusion in situation assessment processes. In: Gabbay, M., Kruse, R., Nonnengart, A., Ohlbach, H.J. (eds.) Qualitative and quantitative practical reasoning, pp. 1–15. Springer, Berlin (1997)
6. Madey, G.R., Blake, M.B., Poellabauer, C., Lu, H., McCune, R.R., Wei, Y.: Applying DDDAS principles to command, control and mission planning for UAV swarms. Procedia Comput. Sci. **9**, 1177–1186 (2012)
7. Endsley, M.R., Bolte, B., Jones, D.G.: Designing for Situation Awareness: An Approach to User-centered Design. FL CRC Press, Boca Raton (2003)
8. Khaleghi, B., Khamis, A., Karray, F.O., Razavi, S.N.: Multisensor data fusion: A review of the state-of-the-art. Inf. fusion. **14**, 28–44 (2013)
9. Zhang, W., Feltner, D., Shirley, J., Swangnetr, M., Kaber, D.: Unmanned aerial vehicle control interface design and cognitive workload: A constrained review and research framework. In: Systems, Man, and Cybernetics (SMC), 2016 IEEE International Conference on, pp. 1821–1826 (2016)
10. McCarley, J.S., Wickens, C.D.: Human Factors Concerns in UAV Flight (2004)
11. Endsley, M.R.: Design and evaluation for situation awareness enhancement. In: Proceedings of the Human Factors Society Annual Meeting, pp. 97–101 (1988)
12. Ruff, H.A., Narayanan, S., Draper, M.H.: Human interaction with levels of automation and decision-aid fidelity in the supervisory control of multiple simulated unmanned air vehicles. Presence Teleoperators Virtual Environ. **11**, 335–351 (2002)
13. Mitchell, H.B.: Multi-sensor Data Fusion: An Introduction. Springer Science & Business Media, Berlin (2007)
14. Bürkle, A., Segor, F., Kollmann, M.: Towards autonomous micro UAV swarms. J. Intell. Robot. Syst. **61**, 339–353 (2011)
15. Pfautz, J., Fouse, A., Fichtl, T., Roth, E., Bisantz, A., Madden, S.: The impact of meta-information on decision-making in intelligence operations. In: Proceedings of the Human Factors and Ergonomics Society Annual Meeting, pp. 214–218 (2005)
16. Nigam, N.: The multiple unmanned air vehicle persistent surveillance problem: A review. Machines **2**, 13–72 (2014)

17. Chen, J.Y.C., Barnes, M.J., Harper-Sciarini, M.: Supervisory control of multiple robots: Human-performance issues and user-interface design. IEEE Trans. Syst. Man Cybern. Part C Appl. Rev. **41**, 435–454 (2011)
18. Lupyan, G., Rakison, D.H., McClelland, J.L.: Language is not just for talking: Redundant labels facilitate learning of novel categories. Psychol. Sci. **18**, 1077–1083 (2007)
19. Gauthier, I., James, T.W., Curby, K.M., Tarr, M.J.: The influence of conceptual knowledge on visual discrimination. Cogn. Neuropsychol. **20**, 507–523 (2003)
20. Giannoukos, S., Brkić, B., Taylor, S., Marshall, A., Verbeck, G.F.: Chemical sniffing instrumentation for security applications. Chem. Rev. **116**, 8146–8172 (2016)
21. Hing, J., Oh, P.Y.: Integrating motion platforms with unmanned aerial vehicles to improve control, train pilots and minimize accidents. In: ASME 2008 International Design Engineering Technical Conferences and Computers and Information in Engineering Conference, pp. 867–875 (2008)

Participant Perception and HMI Preferences During Simulated AV Disengagements

Syeda Rizvi[1(✉)], Francesca Favaro[1,2], and Sumaid Mahmood[1]

[1] RiSAS Research Center, San Jose State University, San Jose, CA, USA
{syeda.rizvi,francesca.favaro,
sumaid.mahmood}@sjsu.edu
[2] Department of Aviation and Technology,
San Jose State University, San Jose, CA, USA

Abstract. This study examined drivers' responses to simulated autonomous technology failures in semi-autonomous vehicles. A population of 40 individuals was tested, considering the following independent variables: age of the driver, speed at disengagement, and time at which the disengagement occurred. Participants received auditory and visual warning at the time of disengagement and were asked to regain control of the vehicle while maneuvering within a S-curve turn. Participants' perception associated to the estimation of success of the control takeover, estimation of test duration, and estimation of the speed of travel showed poor accuracy. Speed recollection accuracy was lower for older participants, while younger participants showed overconfidence in the assessment of the quality of their control takeover. The employed human-machine interface highlighted concerns on the use of central console displays. Trust in the technology and nervousness to the possibility of a disengagement showed higher levels of anxiety for high speeds.

Keywords: Human factors · Human machine interfaces · Driving simulations · Age-related issues · Trust in automation · Physiological measures

1 Introduction

Autonomous Vehicle (AV) technology is quickly expanding its market. Manufacturers are targeting different levels of autonomy, with semi-autonomous vehicles currently in the lead. In semi-autonomous vehicles, a human driver collaborates with the software that acts as "brain" of the vehicle and serves as back-up whenever the Autonomous Technology (AT) disengages after a failure. Current regulations in California for SAE Level 3 semi-AVs (see [5]) require the human driver to "actively monitor the operations of the vehicle," and to be "capable of taking over immediate physical control" [2]. In the safety-critical situation of an AT disengagement, it is important to ensure that the human driver has enough time to react and respond effectively to the request to control the vehicle.

To test such situation, this study simulated an autonomous technology disengagement scenario within a NHTSA-compliant high-fidelity simulator. The simulator was capable of handling both manual/conventional driving, as well as autonomous

J. Chen (Ed.): AHFE 2019, AISC 962, pp. 36–45, 2020.
https://doi.org/10.1007/978-3-030-20467-9_4

driving and works with a human-in-the-loop setting. The study evaluated a number of dependent variables, including reaction times to disengagements, drift performance, and several metrics to quantify human factors associated with the simulation experiment. This paper focuses on this last portion of the results and presents an overview of participants' situational awareness associated to the estimated success of the control takeover (measured as a function of unintentional lane departures), estimation of the time within the simulation, and estimation of the speed of travel. Furthermore, the paper presents highlights collected from surveys that were targeting the preferences of the participants regarding the human-machine interfaces (HMI) as well as subjective measures such as emotional and physical response of the participants to the simulation.

2 Methodology

The study employed a static driving simulator consisting of a BMW 6 series, a projection wall providing 220-degree horizontal front view and a split rear projection wall providing the projection for side and rear-view mirrors. A population of 40 individuals was tested, equally split among male and female, reflecting gender and age distribution of US licensed drivers as reported by the Federal Highway Administration [6].

The study employed a specific scenario of an unstructured AT disengagement, i.e. without prior advanced warning. The participants sat in the vehicle driving in autonomous mode for a predetermined amount of time within a simulated highway environment. At the determined time of disengagement, a visual and auditory warning prompted the human driver to regain manual control of the vehicle. No external visual cues were provided to the driver to let him/her know of the upcoming disengagement (e.g., no road construction sites, no visible obstacles). This was done to test a worst-case scenario that simulated a software-related failure of the AT. This point is something that separates the present study from similar studies that were executed in the past [4]. No distractor was employed and, following California (CA) regulations, all participants were instructed to monitor the vehicle's operation at all times and, should a disengagement happen, to regain control of the vehicle to the best of their ability while remaining within the same lane of travel that the AT disengagement had happened in.

The tests were executed in a 12.23 km closed-track highway-like simulated environment. The track was a combination of four identical-in-shape sections connected by four S-shaped curves. The specific S-curve shape was chosen to measure and test the quality of the control takeback (with respect to the quantitative drift and driving offset metrics analyzed in [3]).

Prior to the beginning of the test, each participant was given the opportunity to train for 5 min on the track in pure manual driving. The actual test began with the car driving autonomously. Three independent variables were selected for the study: (i) age of the driver, with three distinct ranges of 18–35; 35–55 and 55+; (ii) speed, with a high setting of 105 km/h (65 mph - CA highway speed limit) and a low speed setting of 88 km/h (55 mph); (iii) time of disengagement after beginning of the test, with three ranges of 0–10 min, 10–20 min, 20–30 min. Two repetitions were executed, leading to 36 test scenarios with 18 male and 18 female participants (10% of the entire population suffered from nausea, leaving 36 usable data points from the original population of 40).

3 Results

The following section presents the results of the study, divided in: (i) situational awareness; (ii) perception of success; (iii) HMI preferences; (iv) subjective measures – emotional and physical response. The dependence of the results on the study variables is assessed for each category.

3.1 Situational Awareness

Definition of Accuracy. In order to analyze and interpret the perception and situational awareness of participants during the simulation, we used a performance measure termed "Accuracy", commonly used in Machine Learning for results with binary outcomes [7]. It is essentially a fraction of the predictions/answers that a model gets correct. Equation (1) shows the calculation used to obtain the fraction.

$$(TP + TN)/(TP + TN + FN + FP). \tag{1}$$

Accuracy is calculated by summing the true responses (true negatives (TN) and true positives (TP)) and dividing them by the total number of responses, including false ones (false negatives (FN) and false positives (FP)). The Accuracy indicator was employed to quantify the quality of the participants recollection of the speed at time of disengagement, as well as their perception of success in control takeover after disengagement, as explained next.

Recollection of Speed at the Time of Disengagement. After the test completion, participants were asked to report the speed of the vehicle at the time of the simulation disengagement and whether their recollection was based on an actual reading from the digital speedometer placed in the vehicle dashboard. Their responses were evaluated to be correct based on a threshold of $\pm$ 2mph. After comparing their numerical answers against the actual speed in their test, we computed accuracy per Eq. (1). Table 1 shows the results for accuracy related to speed perception for the whole population, with an overall value of 55.5%.

Table 1. Overall accuracy for speed recollection

	Occurrence
TP	18
TN	2
FN	2
FP	14
Accuracy	55.5%

Figure 1 shows the breakdown of the accuracy computation as a function of the investigated factors. Age did not show a statistically significant dependence, while speed of the vehicle indicates a better performance at low speeds. Accuracy was also lower for medium durations of engagement.

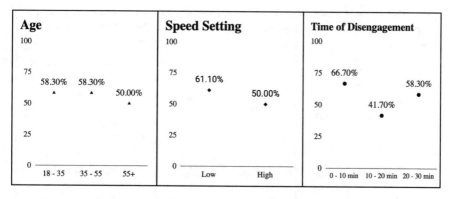

Fig. 1. Accuracy in speed recollection as a function of the investigated factors.

Recollection of Time in the Simulation. Participants were also asked to estimate the duration of the simulation. Their reported times were compared to the actual simulation time by using a threshold of ± 5 min. Additionally, responses were classified as 'Spot On' if their answers had a difference of < 1 min from the actual time. The overall test results are summarized in Table 2, indicating that participants were more likely to overestimate the test duration. Factors dependence showed that the middle-aged group had the higher within threshold estimation percentage, and shorter test durations led to less precise estimations. Figure 2 shows the breakdown of the accuracy computation as a function of the investigated factors. Higher speeds and the middle age group showed a better recollection, whereas the shorter time duration had the lowest accuracy in time recollection.

Table 2. Summary of results for time in simulation

Answers	Occurrence	Percentage (%)	Average error [min]
Underestimated	7	19.4	−4.02
Overestimated	26	72.2	7.09
Spot on	3	8.3	0
Within threshold	**17**	**47.2**	N/A

Recollection of First Input at Disengagement. Participants were asked what their first input provided was as a response to the disengagement. Steering was always the first input and we thus analyzed the accuracy in terms of seeing which participants correctly assessed whether throttle or braking was employed first. Only 22% of the participants actually resorted to braking before throttle as a response to the disengagement and to the change in road curvature. Of the 78% that resorted to acceleration first, 32% (of them) incorrectly recollected braking to be their response to the disengagement event (i.e., they thought they braked but they accelerated instead). This is an important factor to consider, given the prominent effect that higher speeds had on increasing drift and the likelihood of unintentional lane departures.

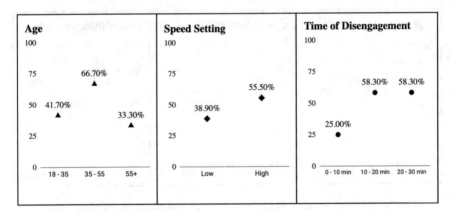

Fig. 2. Percentage of participants within the threshold in time recollection as a function of investigated factors

Gaze Before Disengagement. Participants were asked to rank from 1 (lowest values) to 5 (highest values) their gaze focus level on different parts of the surrounding environment, i.e. outside the vehicle, inside the vehicle and other locations. The overall gaze at different parts of the simulation are represented as weighted averages, shown in Fig. 3. Overall, areas outside the vehicle had the highest gaze levels with the front being the highest, (75% of the participants ranked their gaze as 4 or 5), and other locations including inside the vehicles had considerably lower levels. This is an important point to consider when designing human-machine interfaces, which we discuss shortly.

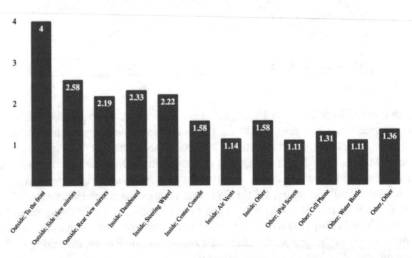

Fig. 3. Participant ranking of gaze focus levels for different locations of the surrounding environment before disengagement shown as weighted averages

3.2   Perception of Success

Perception of success of control takeover. Participants' perception of success in control takeover was studied by comparing a binary option of success (yes/no) indicated by participants to a binary measure of drift (remained within the lane vs. unintentional lane departure). Accuracy is shown in Table 3, indicating overconfidence in the quality of the control takeover by the majority of participants (only one participant recognized the failed recovery attempt).

Table 3. Overall accuracy for recovery success

	Occurrence
TP	11
TN	1
FN	0
FP	24
Accuracy	33.3%

Figure 4 shows again the breakdown as a function of the investigated factors. Improved accuracy was observed for older participants, lower speed settings, and short durations of engagement (which tended to have lower drift). Older aged participants had a higher accuracy (50% vs 25%) in recalling their recovery after the disengagement. Low speeds and lower test durations also led to a better recollection accuracy (50% and 58.3% respectively).

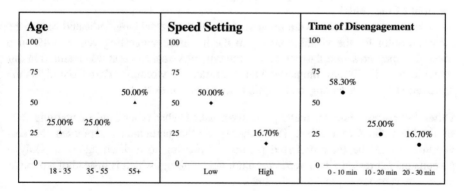

Fig. 4. Accuracy in success estimation as a function of the investigated factors.

3.3   HMI Preferences

An aural warning repeating the phrase "Danger! Take back control" (human male voice) was provided to the participants until they managed to regain control of the vehicle. HMI literature suggested that the word "danger" would create a peak of alertness compared to the word "warning", and that male voice had to be preferred [1].

The visual warning displayed an exclamation point within a chartreuse yellow triangle as well as a symbol of hands on steering wheel. Post-test surveys queried participants' preferences on the interfaces employed. The results highlighted some differences with what literature had suggested. Additionally, the majority of the participants (77.8%) indicated a preference for additional warning though the use of vibration, either of the seat (53%) or of the steering wheel (24.8%).

Aural Warning. 91% of participants reported that they found the aural warning helpful, however 16.7% found it distracting and one participant said it hindered their ability to take control. 80% of the participants indicated that they preferred a human voice, which agreed with what literature pointed. Contrary to literature, the majority (91.2%) of participants indicated that they either preferred a female voice (22.2%) or were neutral on the gender selection (69%).

Visual Warning. A range of experience and preference questions were asked about the visual warning interface. The most significant to note was that 50% of the participants reported not seeing the visual warning provided within the central console (10.2-inch display). Out of the participants that saw it, 2 reported it to be distracting, but that it did not hindered their ability to regain control. 64% of the participants indicated red as the preferred color for the visual, instead of the literature suggested yellow. 64% also stated that they preferred flashing text as the type of visual (no text was featured in our visual icon), with 78% of the participants indicated the wording "Take Control" as the preferred suggestion. Participants also expressed a warning location preference for HUD/windshield displays (36%) or dashboard (30%). This shows that the majority of participants prefer a visual warning that is directly in front of them. This agrees with the gaze levels expressed by participants in Fig. 3, where 75% expressed high rankings to the front of the vehicle.

Additional preferences were queried: 64% of the participants indicated red as the preferred color for the visual, contrary to the literature suggesting yellow. 64% also stated that they preferred flashing text as the type of visual (no text was featured in our visual icon), with 78% of the participants indicated the wording "Take Control" as the preferred suggestion among other options that were presented.

Other Warnings. Finally, participants were asked what other kind of warning they would have liked to receive. The majority of the participants expressed the seat vibration would be their preferred method of alerting to a disengagement. Only 3 participants expressed a preference for automatic braking, which is instead of a popular in current vehicles deployed in the market.

3.4   Subjective Measures – Emotional and Physical Response

The surveys included queries on the participants' emotional state during the simulation. Figures 5 and 6 summarize the key findings in relation to trust, fear, nervousness and anticipation, as well as changes of trust in the technology after the participants' involvement with the study. Fear and nervousness levels were higher at high speeds than low speed. Low speed also led to higher levels of trust. Nervousness was higher for the older age group than the younger one. Moreover, the older age group showed a greater change in trust level than younger ages. Similarly, high speeds showed greater changes than low speed, with a higher percentage of decrease in trust at higher speed than low speed. Longer test duration also showed greater changes than shorter test durations, and there was no decrease in trust for shorter durations. Older age groups, high speeds and longer test durations showed higher levels of anticipation than younger age groups, low speeds and shorter tests.

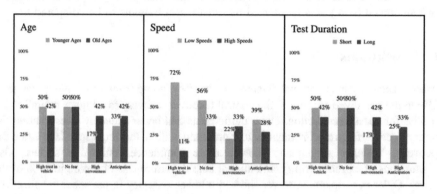

Fig. 5. Summary of main findings related to trust, fear and nervousness and anticipation

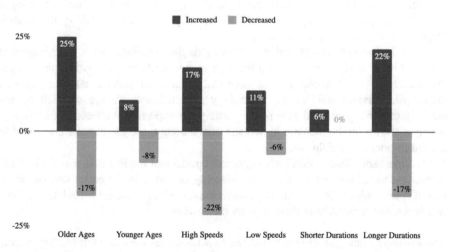

Fig. 6. Changes in level of trust in the AV technology as a function of the investigated speed, age groups and test duration

With regards to physical effects, participants were asked to rate how much they felt nauseous during the simulation, and whether nausea affected their ability to take control of the vehicle following the disengagement. 94% of the participants felt some level of nausea during the test, ranging from mild to moderate. 31% of the participants expressed that nausea affected their driving. 81% of the people who said it somewhat affected their driving did not successfully recover from the disengagement. Moreover, the participant who expressed that it affected their driving significantly had a maximum lane drift in the higher range (3.13 m) while the average drift of all the participants who expressed that nausea affected their driving was 1.59 m, which was slightly higher than the average maximum drift of all participants (1.45 m). Nausea is known to possibly create bias in the results of simulated studies, and this is one of the reasons why a novel metric of the integral offset ratio was set up, so that each participant would have their own simulated (and possibly affected by queasiness) baseline to be compared to.

4   Conclusions

Overall, accuracy in situational awareness was poor, going from a low overall value of 33% in the perceived success of the control takeover recovery to a higher overall value of 55% for speed recollection. Older participants had better accuracy in estimating the binary success/failure with regards to remaining within the requested lane after the recovery. Younger participants showed more confidence, estimating success also in situations of extreme drift (unintentional lane departure was observed in 69% of the cases). Older participants however showed a lower accuracy in the speed recollection. Speed was found to be the most significant factor that affected the results. Speed recollection improved for low speed settings and short durations of engagement, while time recollection was best for long durations of engagement. Recollection of first input also showed poor accuracy, with 32% incorrectly recollecting braking to be their response to the disengagement event.

Trust in the technology and nervousness to the possibility of a disengagement showed higher levels of anticipation for high speeds, and older age groups and longer test durations, while there was an observed balance of participants that reported increase/decrease of trust in the technology after the test. Nausea was felt by the majority of participants and 31% of the participants expressed it affected their driving, which is shown by the increase in average drift among those participants compared to the entire pool of participants.

A remarkable observation with important operational implications was 50% of the population tested did not see the visual warning provided in the center console 10.2-inch display; indeed, 76% of the population expressed a preference for HUD displays, which are just now making their way on the market.

Acknowledgements. The authors would like to acknowledge the help from Mr. Sky Eurich, Mrs. Nazanin Nader, and Mrs. Shivangi Agarwal for the data collection process. The contents of this report reflect the views of the authors, who are responsible for the facts and the accuracy of the information presented herein. The report is funded by a grant from the U.S. Department of Transportation's University Transportation Centers Program (69A3551747127) managed by the Mineta Transportation Institute of San Jose. The U.S. Government assumes no liability for the contents or use thereof.

References

1. Bazilinskyy, P., De Winter, J.C.F.: Analyzing crowdsourced ratings of speech-based take-over requests for automated driving. Appl. Ergon. **64**, 56–64 (2017)
2. California Department of Motor Vehicles (CA DMV).: Article 3.7 – Autonomous Vehicles. Title 13, Division 1, par. 227.32, Autonomous Vehicles Order to Adopt (2018)
3. Favaro, F.M., Seewald, P., Scholtes, M., Eurich, S.: Quality of control takeover following disengagements in semi-autonomous vehicles. Submitted to J. of Transp. Res., November (2018)
4. Mok, B.K.J., Johns, M., Lee, K.J., Ive, H.P., Miller, D., Ju, W.: Timing of unstructured transitions of control in automated driving. In: Intelligent Vehicles Symposium (IV), pp. 1167–1172. IEEE (2015)
5. Society of Automotive Engineers – SAE.: On-Road Automated Vehicle Standards committee. Taxonomy and definitions for terms related to on-road motor vehicle automated driving systems (2014)
6. US Department of Transportation, Federal Highway Administration (FHWA).: Licensed Drivers in the US Demographics for 2016 (2016)
7. Zhu, W., Zeng, N., Wang, N.: Sensitivity, specificity, accuracy, associated confidence interval and ROC analysis with practical SAS implementations. NESUG proceedings: health care and life sciences, Baltimore, Maryland, 19, 67 (2010)

HRI Applications in the Workplace and Exoskeletons

Development of a Standardized Ergonomic Assessment Methodology for Exoskeletons Using Both Subjective and Objective Measurement Techniques

Michael Hefferle[1,2(✉)], Maria Lechner[1,3], Karsten Kluth[2], and Marc Christian[1]

[1] Department for Occupational Safety and Ergonomics, BMW AG,
Moosacherstraße 51, 80809 Munich, Germany
{michael.hefferle,marc.christian}@bmw.de
[2] Ergonomics Division, University of Siegen,
Paul-Bonatz-Straße 9-11, 57068 Siegen, Germany
k.kluth@aws.mb.uni-siegen.de
[3] TUM Department of Sport and Health Sciences, Technical University
of Munich, Georg-Brauchle-Ring 60/62, 80992 Munich, Germany
maria.lechner@tum.de

Abstract. Awkward postures, high loads, and highly repetitive tasks are risk factors for developing work-related musculoskeletal disorders, which are the main reason for sick days in manufacturing. Overhead work, specifically, is a high-risk factor for developing musculoskeletal disorders of the shoulder, which account for the longest sick leaves among all musculoskeletal disorders.

Assistive devices, such as exoskeletons, seek to reduce the stresses associated with overhead work, and have even been suggested as a preventative measure for musculoskeletal disorders.

To investigate the physiological consequences of passive upper limb exoskeletons a standardized holistic assessment methodology, including one subjective (Borg CR-10) and three objective measurement techniques (EMG, ergo spirometry combined with heart rate and NIRS). A set of static, dynamic, and simulated assembly tasks in combination with a suitable test rig is developed and preliminary study results are presented.

Keywords: Overhead work · Work related musculoskeletal disorders · Exoskeleton · Ergonomic assessment · EMG · NIRS · Ergo spirometry

1 Introduction

Musculoskeletal disorders are the main reason for sick days in the automotive industry [1]. Work related diseases such as bursitis, tendinitis, tendon tears, impingement, instability, and arthritis typically, but not exclusively, affect the upper limbs and the shoulder [2]. Prolonged overhead work especially in cumbersome postures are risk

© Springer Nature Switzerland AG 2020
J. Chen (Ed.): AHFE 2019, AISC 962, pp. 49–59, 2020.
https://doi.org/10.1007/978-3-030-20467-9_5

factors for developing diseases of the shoulder, which account for the longest sick leaves among all work related musculoskeletal disorders [2, 3].

Exoskeletons have been introduced as a type of human-robot collaboration where the wearer is in physical contact with the device, allowing for a direct exchange of mechanical power. The devices assist the operator during tasks with poor ergonomics and have been suggested as a preventative measure for musculoskeletal disorders [4]. However, the impact on the physiology of the wearer is not yet fully understood. Hence, it is an ongoing topic of investigation [5]. Looze et al. developed an overview of the conducted ergonomical evaluation studies before 2016. So far, the majority of the conducted studies which focus on industrial applications of exoskeletons primarily investigate the impact on muscle activation measured with electromyography (EMG) of distinct muscle groups [6].

Theurel et al. recently studied the physiological consequences of using a passive upper limb exoskeleton during manual tasks by measuring muscle activation, heart rate and postural balance as well as perceived exertion. Four female and four male participants performed three different tasks, walking, lifting and stacking in a laboratory environment. Positive effects, such as reduced workload on shoulder flexor muscles during load lifting and stacking tasks were accompanied by increased antagonist muscle activity, higher heart rates, postural strains and changes in upper limb kinematics [7].

Weston et al. investigated the changes to the biomechanical loading on the lower back while wearing a passive exoskeletal intervention to support the upper limbs. In a controlled laboratory study, twelve male participants performed an assembly task – screw driving in an overhead and a frontal posture – while performing electromyographic measurements of the back and the torso. Using motion capturing data and a force plate, spinal loads were assessed by running a simulation. The results showed an increase in both peak and mean muscle forces in the torso extensor muscle as well as an increase in spinal load compression [8].

Kim et al., Wu et al., Huysamen et al. and Muramatsu et al. also used electromyography to assess local muscle activation of various muscle groups while wearing both active and passive exoskeletons to support the upper limbs during static holding and assembly tasks. Electromyographic measurements were complemented by subjective evaluation of perceived discomfort and/or measurements of functional performance and postural balance [9–13].

All of the above-mentioned investigations measured the muscle activation of different muscle groups to assess the physiological impact of exoskeletons on the wearer. In most cases, only muscle groups that are activated due to performing a certain movement, which is supposed to be augmented by the exoskeleton device, were investigated. A few researchers also analyzed the impact on either the antagonistic muscle group to the augmented muscles, or muscle groups they assumed to have different activation levels due to an altered posture, movement pattern or weight compensation.

The choice which muscles are measured inherently contains a certain bias to the presumed results and may disregard the possibility that augmenting specific movements, e.g. muscles through an exoskeleton, could very well have a far greater effect on human physiology than the muscles that are directly or indirectly involved [14]. Since it is neither practical nor feasible to use electromyography to assess the impact on each muscle that might be affected – which could be in theory every single muscle – a holistic approach as proposed earlier by Dahmen and Hefferle [15] should be considered.

This paper presents a holistic approach for a standardized ergonomic assessment methodology for passive upper body exoskeleton devices. Both subjective and objective measurement techniques are implemented and an experimental design in addition to a suitable test rig is presented.

2   Methods

2.1   Exoskeleton Device

Passive upper limb exoskeleton devices support the wearer by compensating the weight of the arms to reduce the stresses on the shoulder during prolonged overhead work (the term commonly refers to tasks where the arms are raised to a level at or above the shoulders). The commercially available devices for industrial applications are worn like a backpack. The upper arms are typically attached to pads, which are connected to the main frame. This frame, a rigid structure on the back, commonly combined with a Bowden cable or spring mechanism redirects the loads, which normally are experienced by the shoulder region, through the exoskeleton frame into the hips. A more detailed functional description of a upper limb exoskeleton has been previously reported [15].

2.2   Experimental Design

Overhead work places in the automotive industry require the worker typically to perform tasks standing in an upright posture while raising the arms at or above shoulder level. Detailed characteristics of an overhead work place at an automotive assembly line have been previously described [15].

To assess the extent of a passive upper limb exoskeleton on human physiology, a set of movements and postures (in this case, a distinct joint angle) is introduced as shown in Fig. 1. Three static postures, three dynamic movement tasks, and two simulated assembly tasks define the experimental conditions.

Control conditions are boldfaced. The postures, movements and tasks were selected to represent the characteristics of overhead work places commonly appearing at an automotive assembly line. Distinguishing between static, dynamic and assembly tasks is also in accordance with the movements or tasks in the investigations, mentioned in the introduction chapter.

For the static task blocks (S1/S4, S2/S5, S3/S6) the angles of the shoulder and the elbow vary between 0°, 90° and 180° degrees. The shoulder angle is measured between the arm and the vertical axis of the body (0° means no arm flexion) and elbow angle between the lower and the upper arm (0° means no forearm flexion). During dynamic tasks (D1/D4, D2/D5, D3/D6) shoulder and elbow angles vary between 0° and 180° degrees at constant speed of 60 or 45 Hz between start and end position. A sound signal using a metronome indicates start and end position, to help participants maintain a constant frequency. For each of the six static and dynamic tasks participants are

control / intervention blocks		Control - No Exo / Intervention - Exo							
tasks blocks		Static			Dynamic			Sim. Assembly	
Conditions		S1/S4	S2/S5	S3/S6	D1/D4	D2/D5	D3/D6	St1/St2	Sc1/Sc2
Variables	Units								
TASK STATIC/DYNAMIC	[%S/%D]	100/0	100/0	100/0	0/100	0/100	0/100	50/50	50/50
CLOCK FREQUENCY	[Hz]	-	-	-	60	60	45	-	-
QUANTITY	[x in/x out]	-	-	-	-	-	-	15 / -	10 / 10
SHOULDER ANGLE	[°]	90	90	180	0-90	90-180	0-180	-	-
ELBOW ANGLE	[°]	0	90	0	0	90-0	0-90-0	-	-
LOADING CONDITION	[kg]	0	0	0	0	0	0	0	3
DURATION	[s]	60	60	60	60	60	60	-	-

Fig. 1. Characteristics of experimental conditions

standing in an upright posture on a spot marked on the ground (see also paragraph 2.3). Tasks are performed for 60 s, which relates to a typical length of a cycle at an automotive assembly line. For the two simulated assembly tasks, plug mounting (St1/St2) and screw driving (Sc1/Sc2), no time limit is given, but the number of plugs to be mounted as well as the amount of screws that have to be driven in and out is specified. Table 1 contains a detailed explanation of the postures, movements or tasks for the experimental conditions.

Table 1. Postures, movements or tasks of the experimental conditions.

Condition	Posture / Movement / Task Description
S1/S4	Arms raised to 90° shoulder flexion, elbow flexion 0°
S2/S5	Arms raised to 90° shoulder flexion, elbow flexion 90°
S3/S6	Arms raised to 180° shoulder flexion, elbow flexion 0°
D1/D4	Start position: Arms neutral, 0° shoulder flexion; elbow flexion 0° End position: Arms raised to 90° shoulder flexion; elbow flexion 0°
D2/D5	Start position: Arms raised to 90° shoulder flexion; elbow flexion 90° End position: Arms raised to 180° shoulder flexion; elbow flexion 0°
D3/D6	Start position: Arms hanging, 0° shoulder flexion; elbow flexion 0° End position: Arms raised to 180° shoulder flexion; elbow flexion 0°
St1/St2	Plug mounting of 3 × 5 plugs (diameter 35, 25, 20 mm); Participant steps forward, retrieves a stack of 5 plugs out of a box attached to the test rig and applies the plugs on the panel. Task is repeated until all stacks are mounted. Start and end position: marked spot on the ground.
Sc1/Sc2	Screw driving in and out, 10 screws (M8); Participant steps forward, retrieves the screw driver from a stool, picks up a single screw out of a box and screws it into the panel. Task is repeated until all screws are screwed in. Participant immediately starts screwing out again after finishing. Once a screw is screwed out subject will grab it with the other hand to maintain an overhead work posture. After 5 screws are screwed out and collected in one hand, subject is allowed to put them back into the box and the process is repeated once more until all 10 screws are put back into the box. After finishing subject puts the screw driver back on the stool. Start and end position: marked spot on the ground.

Characteristic postures or movements form an experimental task block. For instance, the three static postures form the static task block. Using a within-subject design, each subject performs all task blocks twice, without and with intervention, i.e. exoskeleton device, which leads to 16 experimental conditions. Conditions are systematically varied and participants are randomly assigned.

In the first step, the control condition ("No Exo") and the intervention condition ("Exo") is varied. In the second step, within the first block i.e. control or intervention condition the order of the three task blocks (static, dynamic, simulated assembly) is varied systematically, while the order of the single tasks within each of the three task blocks is not varied. In the third step, the second control or intervention block, the order of the three task blocks is mirrored. Participants will start with the condition block with which they finished the first control or intervention condition. Having the constraint of the mirroring pattern 12 different experimental conditions are possible.

Figure 2 shows a diagram of the variation systematization.

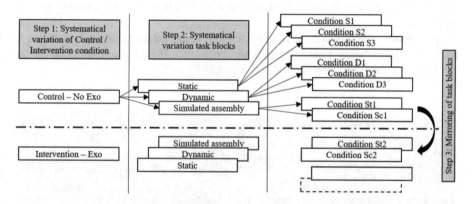

Fig. 2. Systematical variation of conditions and task blocks.

Complete randomization of all 16 experimental conditions is rejected. Randomizing the task blocks, but not the tasks themselves, as well as mirroring the task blocks between control and intervention condition is chosen due to three advantages:

1. If randomization doesn't occur between control and intervention conditions the risk of sensor disruption during the donning and doffing of the exoskeleton is minimized.
2. Participants are less easily confused, when the order of the single tasks do not vary within a task block.
3. Mirroring the task blocks may allow investigating whether participants' subjective assessments are affected differently if the same tasks between control and intervention condition are compared immediately or after performing several different tasks in between.

Before the experiment starts, MVCs (maximum voluntary contraction) of the muscle groups are performed to normalize EMG and NIRS (explanation see paragraph 2.4) data. Each experimental block then starts with the subject resting, seated, for five minutes while the cardio pulmonal base line is measured.

Between the two experimental blocks, control and intervention, trials are paused and participants are allowed to drink and leave the test rig if necessary. Data recording is stopped and resumed later.

2.3 Laboratory Setup

The test rig consists of a 2,00 m × 1,50 m × 2,50 m (length × width × height) rod system with a height adjustable panel (between 1,50 m and 2,50 m) for the simulated assembly tasks. For the base line measurements and resting between experimental conditions, participants are sitting on a chair. A screen attached to the test rig, which provides participants with detailed instructions on the subsequent posture or task for each experimental condition, is in direct line of sight when adopting a resting position on the chair.

Participants completed subjective evaluations after each experimental condition by filling in a paper-based questionnaire while standing at the table to the left of the test rig. Figure 3 shows the test rig and the adjustable panel.

Fig. 3. Test rig with marked spot, screen and bar table (left) and adjustable panel for simulated assembly tasks (right)

Static postures and dynamic tasks are performed standing in an upright posture on a marked spot on the ground in the middle of the test rig to guarantee equal conditions for each subject. Panel height (H_P) is adjusted according to individual body dimensions of the participants, applying the following formula:

$$H_P = H_S + L_{ES}. \tag{1}$$

The height of the shoulder (H_S) is measured from the ground to the acromion of the shoulder while wearing footgear and standing upright against a wall. To measure the length between the elbow and the end of the screwdriver pit (L_{ES}), participants are asked to hold the screwdriver in their dominant hand while posing in the standard overhead work posture (90° shoulder flexion between the vertical axis and the arm; 90° elbow flexion between upper and lower arm; wrist straight). Assumed joint angles are varified with a goniometer prior to measurement.

2.4 Data Collection

For data collection, four different techniques (see Fig. 4), one subjective – a questionnaire including a Body-Map, and three objective – EMG, NIRS and ergospirometry including heart rate, are deployed.

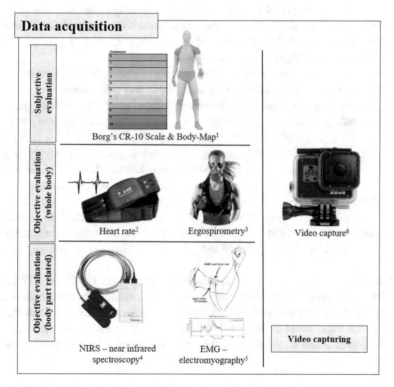

Fig. 4. Methods for data acquisition ([1]*BMW AG*, [2]www.polar.com/de, [3]www.procarebv.nl, [4]www.artenis.nl/com, [5]www.biomechanik.sg.tum.de/labor/elektromyographie, [6]gopro.com)

Subjective evaluation of perceived effort on 18 different body parts – including lower and upper torso, upper limbs, shoulder and neck – and the whole body is assessed using Borg's CR-10 Scale [16] combined with a modified Body-Map after Corlett and Bishop [17].

Muscle activation of both left and right upper trapezoid and lateral deltoid muscles is measured with surface EMG (myon 320, myon AG, Schwarzenberg, Switzerland) to assess differences in stress and fatigue between control and intervention condition. Placement of the electrodes strictly follows SENIAM guidelines as suggested previously [14].

Near-infrared spectroscopy (NIRS) (PortaMon, Artinis Medical Systems, Zetten, The Netherlands) is a noninvasive optical measurement method to quantify blood oxygenation levels. It is based on changes in the near-infrared absorption characteristics of hemoglobin (Hb) in local tissue, which differs from the level of oxygen bent [18]. O_2Hb (blood oxygenation) and tHb (total blood volume) are measured to assess the oxygenation status and hemodynamics of the anterior deltoid muscle on both sides. Muscle fatigue will be assessed as described by Ferrari et al. [19] and the method will be adapted for an ergonomic exoskeleton assessment as previously reported by Muramatsu et al. [18]. NIRS is also implemented to investigate whether performing several tasks in an overhead posture of the upper limbs using an exoskeleton device can be associated with blood flow disorders of the shoulder muscles, which may indicate possible negative side effects of an exoskeleton device.

A synchronization remote control (PortaSync, Artinis Medical Systems, Zetten, The Netherlands) is used to synchronize both EMG and NIRS data.

To estimate the physiological consequences without any further bias as discussed in the introduction paragraph, ergospirometry is used to objectively quantify the impact on the whole body. Oxygen consumption via respiratory measurements (METAMAX 3B-R2, Cortex Biophysik GmbH, Leipzig, Germany) is used to estimate the metabolic effort and has been proposed earlier to evaluate exoskeleton devices [14]. Participants wear a mask that samples the oxygen and carbon dioxide content of each individual breath, from which energy expenditure is calcuated [20]. Since heart rate sensors commonly accompany ergospirometrical systems, it is not counted as an additional measurement technique. Heart rate is continuously monitored using a chest strap with a built-in sensor (Polar Team, Polar Electro, Kempele, Finland) to estimate metabolic effort as well.

Since oximetry has a very low sampling rate compared to EMG and NIRS and the delay between physical exercise and measurable changes in the cardiopulmonal readings is significantly delayed [20], manual synchronization is found to be sufficient.

The entire experiment is captured on video to control joint angles and movements of the participants and may help to discover possible errors in the recorded data in a later stage.

3   Results

Four participants, three male and one female (Age 26.3 ± 3.5 years, statue: 177.1 ± 5.8 cm, weight: 69.7 ± 10.7 kg), participated in preliminary investigations in order to evaluate and improve the experimental design.

Two additional body parts – the anterior shoulder region; left and right – were added to the Body-Map because participants reported different stress levels between the anterior and the posterior shoulder region.

During preliminary tests, heart rate readings were either implausible or the sensor unit entirely stopped working, while filling in the questionnaire. Sitting in a forward bent posture, the chest strap might have detached slightly and the contact between the wearer and the strap was lost. The regular table was replaced by the bar table where participants are standing while filling in the questionnaire. Additionally, electrode gel was applied between the skin and the chest strap to ensure best possible contact, which resolved the above stated problem.

Baseline measurements and resting periods initially took place in a standing posture, which resulted in unstable, unnaturally high, cardiopulmonary readings, due to small movements, e.g. shifting the body weight from one foot to another. Hence, baseline measurements and resting periods are performed while sitting on a chair to address this issue.

Resting times between the experimental conditions had to be increased from 90 to 120 s, allowing participants' metabolism enough recovery time for their heart rate and oxygen intake to assume base line values again.

4 Discussion

The average age of the four participants who took part in the preliminary tests is not representative for the average work force at the assembly line, as they were considerably younger. Age may have an influence on the recovery times of the metabolism and the suggested 120 s resting times may need to be increased.

Additionally, participants' skill level did not measure up to the skill level of trained workers, which might have had an influence on how tasks are performed between control and intervention condition. The influence might be greater on the results of the simulated assembly tasks, since movements are less constraint and the training effect between the individual trials is more evident.

Within the systematical variation scheme, mirroring on the task-block level but not on the task levels is not true mirroring. For the given example in Fig. 2 true mirroring would mean that the conditions St1 and Sc1 would follow conditions Sc2 and St2 instead of St2 and Sc2. The effect that participants' subjective assessments are affected differently might not be as apparent as desired and the interpretation of future results must therefore be either treated with caution, or true mirroring should be applied.

5 Conclusion

This paper reviews the previously conducted ergonomic assessment studies on exoskeleton devices, which revealed that most of the previous studies focused their EMG measurements on the augmented muscle groups or other distinct muscles groups (e.g. antagonistic muscles). The selection of specific muscle groups to measure is inherently biased. Therefore, a holistic approach, which includes one subjective and three objective measurements techniques, is proposed. An experimental design which allows the deployment of the aforementioned measurement techniques with a suitable laboratory setup is developed. Finally, results of preliminary investigation are presented and discussed.

References

1. Statistical Office of the European Communities.: Health and Safety at Work in Europe (1999–2007). A Statistical Portrait. Eurostat. Statistical Books. Office for Official Publications of the European Union, Luxembourg (2010)
2. American Society of Biomechanics (Hrsg).: EMG assessment of a should support Exoskeleton during on-site job tasks (2017)
3. Bargende, M., Reuss, H-C., Wiedemann, J. (Hrsg).: 17 Internationales Stuttgarter Symposium. Automobil- und Motorentechnik. Proceedings. Springer Fachmedien Wiesbaden, Wiesbaden (2017)
4. Spada, S., Ghibaudo, L., Gilotta, S., Gastaldi, L., Cavatorta, M.P.: Investigation into the applicability of a passive upper-limb exoskeleton in automotive industry. Procedia Manufacturing **11**, 1255–1262 (2017). https://doi.org/10.1016/j.promfg.2017.07.252
5. Spada, S., Ghibaudo, L., Gilotta, S., Gastaldi, L., Cavatorta, M.P.: Analysis of exoskeleton introduction in industrial reality: main issues and EAWS risk assessment. In: Goonetilleke, R.S., Karwowski, W., 0009835 (Hrsg) Advances in Physical Ergonomics and Human Factors. Proceedings of the AHFE 2017 International Conference on Physical Ergonomics and Human Factors, July 17–21, 2017, The Westin Bonaventure Hotel, Los Angeles, California, USA. Springer International Publishing, Cham, s.l (2018)
6. de Looze, M.P., Bosch, T., Krause, F., Stadler, K.S., O'Sullivan, L.W.: Exoskeletons for industrial application and their potential effects on physical work load. Ergonomics **59**(5), 671–681 (2016). https://doi.org/10.1080/00140139.2015.1081988
7. Theurel, J., Desbrosses, K., Roux, T., Savescu, A.: Physiological consequences of using an upper limb exoskeleton during manual handling tasks. Appl. Ergon. **67**, 211–217 (2018). https://doi.org/10.1016/j.apergo.2017.10.008
8. Weston, E.B., Alizadeh, M., Knapik, G.G., Wang, X., Marras, W.S.: Biomechanical evaluation of exoskeleton use on loading of the lumbar spine. Appl. Ergon. **68**, 101–108 (2018). https://doi.org/10.1016/j.apergo.2017.11.006
9. Huysamen, K., Bosch, T., de Looze, M., Stadler, K.S., Graf, E., O'Sullivan, L.W.: Evaluation of a passive exoskeleton for static upper limb activities. Appl. Ergon. **70**, 148–155 (2018). https://doi.org/10.1016/j.apergo.2018.02.009
10. Kim, S., Nussbaum, M.A., Mokhlespour Esfahani, M.I., Alemi, M.M., Alabdulkarim, S., Rashedi, E.: Assessing the influence of a passive, upper extremity exoskeletal vest for tasks requiring arm elevation. Part I – "Expected" effects on discomfort, shoulder muscle activity, and work task performance. Appl. Ergon. (2018). https://doi.org/10.1016/j.apergo.2018.02.025
11. Kim, S., Nussbaum, M.A., Mokhlespour Esfahani, M.I., Alemi, M.M., Jia, B., Rashedi, E.: Assessing the influence of a passive, upper extremity exoskeletal vest for tasks requiring arm elevation. Part II – "Unexpected" effects on shoulder motion, balance, and spine loading. Appl. Ergon. (2018). https://doi.org/10.1016/j.apergo.2018.02.024
12. Wu, W., Fong, J., Crocher, V., Lee, P.V.S., Oetomo, D., Tan, Y., Ackland, D.C.: Modulation of shoulder muscle and joint function using a powered upper-limb exoskeleton. J. Biomech. **72**, 7–16 (2018). https://doi.org/10.1016/j.jbiomech.2018.02.019
13. Muramatsu, Y., Kobayashi, H., Sato, Y., Jiaou, H., Hashimoto, T., Kobayashi, H.: Quantitative performance analysis of exoskeleton augmenting devices - muscle suit - for manual worker. Int. J. Autom. Technol. **5**(4), 559–567 (2011). https://doi.org/10.20965/ijat.2011.p0559
14. Lazzaroni, M., Toxiri, S., Ortiz, J., De Momi, E., Caldwell, D.G.: Towards standards for the evaluation of active back-support exoskeletons to assist lifting task

15. Dahmen, C., Hefferle, M.: Application of Ergonomic Assessment Methods on an Exoskeleton Centered Workplace Proceedings of the The XXXth Annual Occupational Ergonomics and Safety Conference
16. Borg, G.: Borg's Perceived Exertion and Pain Scales. Human Kinetics, Champaign, Ill (1998)
17. Corlett, E.N., Bishop, R.P.: A technique for assessing postural discomfort. Ergonomics **19**(2), 175–182 (1976). https://doi.org/10.1080/00140137608931530
18. Muramatsu, Y., Kobayashi, H.: Assessment of local muscle fatigue by NIRS - development and evaluation of muscle suit -. Robomech J. **1**(1), 46 (2014). https://doi.org/10.1186/s40648-014-0019-2
19. Ferrari, M., Mottola, L., Quaresima, V.: Principles, techniques, and limitations of near infrared spectroscopy. Can. J. Appl. Physiol. **29**(4), 463–487 (2004)
20. Koller, J.R., Gates, D.H., Ferris, D.P., David Remy, C.: 'Body-in-the-loop' optimization of assistive robotic devices: a validation study. In: Hsu, D., Amato, N., Berman, S., Jacobs, S. (Hrsg) Robotics: Science and Systems XII. Robotics Science and Systems Foundation, Berlin? (2016)

Human-Robot Cooperation in Manual Assembly – Interaction Concepts for the Future Workplace

Henning Petruck[1(✉)], Jochen Nelles[1], Marco Faber[1], Heiner Giese[2],
Marius Geibel[2], Stefan Mostert[2], Alexander Mertens[1,3],
Christopher Brandl[1,3], and Verena Nitsch[1]

[1] Institute of Industrial Engineering and Ergonomics of RWTH
Aachen University, Bergdriesch 27, 52062 Aachen, Germany
{h.petruck, j.nelles, m.faber, a.mertens,
c.brandl, v.nitsch}@iaw.rwth-aachen.de
[2] Item Industrietechnik GmbH, Friedenstraße 107-109,
42699 Solingen, Germany
{h.giese, m.geibel, s.mostert}@item24.com
[3] ACE – Aachen Consulting for Applied Industrial Engineering
and Ergonomics UG, Im Mittelfeld 91, 52074 Aachen, Germany

Abstract. A human-robot cooperation workstation was developed and implemented as a platform for the examination of ergonomic design approaches and human-robot interaction in manual assembly. Various control modalities are being tested for this workstation, which enable a broad range of applications for human-robot interaction and control. These modalities include computer-generated control commands, gesture-based control using Myo Armbands, force-sensitive control by guiding the robot, motion tracking of the operator, and head-based gesture control using an Inertial Measurement Unit (IMU). The focus is on human-centered and ergonomic development of interaction patterns for these control modalities. This paper presents the multimodal interaction concept with the robot and allocates the presented modalities to suitable application areas.

Keywords: Human-robot interaction · Ergonomic workplace design · Occupational safety

1 Introduction

Due to shorter product life cycles and constantly growing product portfolios, highly automated assembly processes reach their limits because automation is no longer economically feasible in all cases. This applies in particular to high-wage countries such as Germany, where the task-specific degree of automation in production is very high. Consequently, there is a need for more flexible production systems that are able to react more dynamically to product changes. One possibility for flexible design in production is the integration of humans in the form of human-robot cooperation (HRC). Thus, a semi-automated production can be supplemented by manual activities in which

© Springer Nature Switzerland AG 2020
J. Chen (Ed.): AHFE 2019, AISC 962, pp. 60–71, 2020.
https://doi.org/10.1007/978-3-030-20467-9_6

the robot assists the human working person. The human working person contributes its cognitive and sensorimotor skills, which are complemented by the precision and fatigue-free work of the robot. Another reason for the use of HRC is the reduction of required space. Compared to conventional industrial robots, HRC does not require a safety zone around the robot if the robot and the sensory safety devices are selected appropriately, or the safety zone is considerably smaller, so that safety cages around the robot are not necessary. This also facilitates the integration of HRC systems into existing manual assembly lines, where there is usually no possibility of enlarging the space of the workstation. In many cases, HRC thus contributes to relieving the strain on humans during repetitive activities, like the assembly of car door seals [1]. In this context, HRC is also often referred to when there is no direct cooperation between human and robot, but only coexistence in the same workspace as, for example, with the adhesive application at Audi [2].

The application of HRC is accompanied by new challenges. In particular, occupational safety plays an important role, which must be ensured in direct interaction with the robot. Standards such as the DIN EN ISO 10218 Part 1 (robots) [3] and Part 2 (robot systems and integration) [4] as well as ISO/TS 15066 [5] already provide a framework which has to be considered when setting up HRC-workstations. Besides occupational safety, the implementation of ergonomic interaction concepts with the robot is also important for the usability of the robot. The robot's movements and actions should be transparent and comprehensible in order to generate user confidence in the new "working partner" [6]. The intuitive design of the interaction and control of the robot, as well as the task allocation between man and robot, represent key factors for successful cooperation. However, there is not "the one" HRC workstation. Due to different, individual application scenarios, the appearance of HRC workplaces is rather characterized by diversity and the interaction concepts are tailored individually to the respective conditions. For this reason, it is not possible to set up one workplace at which all HRC-related research questions can be examined holistically, but there is still a need for flexible HRC platforms allowing as many questions as possible to be examined.

The "Collaborative Workplace for Assembly" (CoWorkAs) – developed and set up at the Institute of Industrial Engineering and Ergonomics (IAW) of the RWTH Aachen University in cooperation with item Industrietechnik GmbH – is intended for use in robot-assisted assembly and is implemented at the IAW as a research platform for investigating a wide variety of research questions relating to HRC. The robot meets the outlined requirements, e.g. through the integrated height adjustment and variable positioning of the robot as well as flexible connection possibilities to conveyor belts and logistics systems. Another part of the flexible workstation concept is a multimodal control concept for the robot, which is presented in this article. The aim is to ensure the accessibility and controllability for various user groups of the workplace equally and at the same time to achieve an ergonomic design. By the implementation of different interaction concepts a broad application spectrum for working with the robot is enabled, which serves as basis for the derivation of generally valid design references of interaction concepts. In order to cover the broadest possible range of applications, interaction concepts should be selected such that they can be used for a high degree of automation on the one hand and for manual human control on the other. In addition, a distinction is made between application scenarios in which there is a divided workspace between human and robot, and applications of HRC without spatial separation. The aim of the contribution is to present the selected interaction modalities and concepts for the control of the robot. They have been selected in such a way that there is at least one control mode for each combination of the listed features degree of automation and workspace division.

2 State of the Art

A variety of interaction concepts is required to control the robot so that a wide range of application scenarios can be implemented and investigated with the platform. Various technologies can be used for this purpose. In the following, selected interaction modalities such as hand gestures and head control as well as measurement techniques and sensors such as electromyography (EMG), time of flight cameras or IMU for a human-robot interaction are presented. The selection of the interaction concepts is intended to achieve a broad coverage of the possible characteristics of both the degree of automation and the division of workspace.

Schmidt et al. [7] describe gesture control, which is both imaginable as a manual control method with or without spatial separation, as an intuitive form of interaction, since it has its origin in human communication and therefore offers the possibility of interacting with the robot in a natural way. In a laboratory study, the authors tested an intuitive 3-D gesture control in comparison to a conventional gamepad control for a mobile service robot to determine the caused stress. The obtained results showed that the gesture control is less suitable for driving tasks of the mobile service robot, because there were many collisions with the road markings, but it is promising for the control of a manipulator arm of the robot. Here the number of collisions with the object of investigation could be reduced compared to the gamepad controller.

Another type of gesture control is the use of Myo Armbands from Thalmic Labs, which can detect both hand gestures via EMG and relative hand movements via an IMU. The detected hand gestures can be used to control the robot's actions [8, 9].

Sanna et al. [10] and Stowers et al. [11] are using the Microsoft Kinect as a depth imaging camera to implement a drone control system that allows the drone to be manually controlled by gestures and tracked arm movements. The control can be used with or without spatial separation. There are several approaches to controlling a robot using the Kinect. In some cases, only gesture-based control has been implemented, where pre-defined control commands are triggered by the operator's hand gestures [12, 13]. In other approaches a position control was implemented, which additionally uses the markerless motion tracking of the Kinect [14, 15]. This approach is also suitable for use in a shared workspace, where the robot can react with a high degree of automation to changes in the human position, i.e. it is only indirectly controlled by humans by slightly adapting the target positions to the human position, and the operator can act freely. Li et al. [16] even combine the techniques Myo-armband and Kinect to control the 2-arm robot Baxter.

In addition, Rudigkeit et al. [17] and Jackowski et al. [18] present a manual control concept for a lightweight robot using a 9-axis IMU. The concept aims at enabling, for example, people with a physical disability such as paraplegia to control a jointed-arm robot using of head movements and head gestures by means of an IMU attached to a hair circlet. The designed control paragdigm uses a graphical user interface to map the seven degrees of freedom of the robot (six degrees of freedom by the jointed-arm robot and one degree of freedom by opening/closing the gripper attached to the end effector) with the three degrees of freedom of the human head. This control concept was investigated by the authors both with paraplegic people [19] and with workers of different age groups with regard to stress and strain during the interaction with the robot [20]. The findings gained in these studies are incorporated into the development of the HRC workplace presented here. Depending on the scenario, this control system can also be used with or without spatial separation.

A widespread method for teaching robot movements is kinesthetic teaching [21, 22]. For example, the robot can be taught the task of ironing or opening the door by demonstrating the needed movements for this task [23]. This type of interaction can only take place in a shared workspace, since a haptic interaction between human and robot is required. After the teaching, the robot can perform the activities autonomously.

Industrial robot controllers, such as the KUKA smartPAD, allow the control of the individual axes of the robot, as well as the movement and rotation of the end effector using a 6-dimensional mouse [24]. This type of operation is often used for teaching, where individual positions are approached and stored. However, it does not allow freehand operation and is therefore not suitable for all applications. After teaching the robot moves automatically and acts with or without spatial separation from the human being, depending on the installed sensors.

3 HRC Research Platform "CoWorkAs"

The HRC research platform CoWorkAs was developed and built on the foundation of the vision of a manual assembly island, which is embedded in a highly automated, multi-variant manufacturing process. In this scenario, the robot assists the worker during assembly by feeding components and tools. The workstation is connected to the automated part of production by a conveyor belt, which transports both incoming and outgoing parts. The workstation is used to carry out assembly steps that cannot be automated economically. This applies especially to the assembly steps, which represent the large number of variants, as the automation cost/benefit ratio is not economical here. An exemplary visualization of the scenario is depicted in Fig. 1.

The workplace can be divided into two work areas. There is a working surface at the front of the workstation. Since the transfer of components between robot and working person also takes place in this area, this can be regarded as a divided working area. The second working area is located behind this work surface inside the workstation, which is framed by four pillars. This area also represents the interface to logistics and contains an intermediate storage area for components and tools. Components arriving via the conveyor belt are sorted into the intermediate storage area and the (semi-)finished products and empty component containers are returned to the conveyor belt. Incoming components and containers are identified by RFID sensors. In contrast to the assembly area, this work area is reserved solely for the robot.

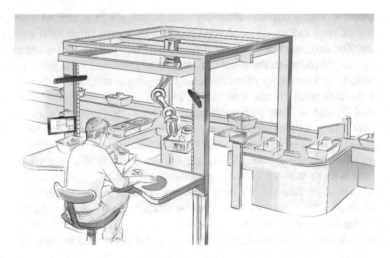

Fig. 1. Visualization of the vision "human-robot collaboration in multi-variant production"

The Powerball Lightweight Arm LWA 4P robot from Schunk used at this workstation is mounted hanging on two linear axes. This type of mounting and the extension of the robot's movement space by two translational axes offer the advantage that both working areas are completely accessible for the robot and at the same time the operator is not restricted in his movement space. Figure 2 shows the actual construction of the workstation.

Fig. 2. Handover of a tool at the HRC research platform "CoWorkAs" (Picture: Ahrens + Steinbach Projekte)

4   Interaction Concept

The vision of a versatile HRC workstation at which different scenarios can be implemented requires a similarly versatile interaction and control concept for the robot. In addition, people in different roles such as the operator or the system integrator must interact with the robot. For example, the system integrator must be able to plan complex trajectories for fetching and delivery movements of the robot with high precision requirements. S/he plans these either without spatial separation using kinesthetic teaching or with spatial separation using an industrial robot control system such as the KUKA smartPAD. For the operator, who usually has only little or no expertise in trajectory planning, an intuitive control option is to be created at the same time, which will enable her/him to control the robot during his work without any major additional effort. The robot usually moves automatically and reacts to human approaches in the divided work area. The idea of inclusion also has an impact on the workplace design. HRC represents a great opportunity to reintegrate physically restricted people into working life. Therefore, the development of control options for these, in particular a head-based gesture control for disabled persons, is also taken into account when implementing the interaction concept of CoWorkAs. In addition, gesture-based control via Myo armbands, force-based control by actively exerting force on the Tool Center Point (TCP) of the robot arm, control by markerless motion capture using the Microsoft Kinect and manipulation by the robot control on the computer are presented. The control modalities are integrated by the software framework Robot Operating System (ROS), which allows an easy change between the different control modes and contains many basic functions for their realization, e.g. collision-free planning of motion trajectories. The individual control modalities are described in detail below.

4.1   Computer-Based Control

In regular operation mode, the robot carries out picking, delivering and placing movements as well as the transfer of components autonomously. A state graph is used to model how the robot reacts to events such as components arriving from the conveyor belt or the completion of an assembly step by the working person. The movements of the robot require exact positioning. Therefore, exact poses for the robot can be specified and applied by the robot using the computer-based control. The flexible design of the controller allows dynamic adaptation to components of different sizes. Using the ROS framework, trajectories are automatically created for the movement between the start and end point of a movement. The computer-based control serves as the basis for the implementation of the other control modalities.

4.2   Head-Based Control via IMU

During the head-based control the robot is controlled by head movements. The necessary control signals are measured by an IMU (FSM-9, Hillcrest Laboratories), which is attached to the head with a hairband. The measuring unit consists of an acceleration, rotation rate and magnetic field sensor as well as a data processor and records the head movements around the X-, Y- and Z-axis.

Initially before the application of the head-based robot control is started, the rest position of the head and the head deflections of the working person in all three degrees of freedom are measured. This serves to adapt the head-based robot control individually to the movement space of the head. Particularly in the case of older workers or people with physical disabilities such as paraplegia, individual head movements can only be used to a limited extent for head-based control [19, 20]. In addition, the control mode considers that the head movements are carried out in a comfortable range for the working person.

The seven degrees of freedom of the robot (including the gripper) are transferred by means of a control paradigm according to Rudigkeit et al. [17] and Jackowski et al. [18] to the three degrees of freedom of the human head ((counter-)clockwise rotation, flexion and extension as well as lateral flexion to the right and left) by means of four control groups (horizontal plane, vertical plane, orientation in space, gripper opening/closing). The change between these four control groups is realized by means of a graphical user interface, which is displayed on a monitor integrated into the HRC workstation. The selection of a control group by the user is carried out either by means of a so-called cursor control, comparable to the control of a mouse cursor on a computer screen, or directly by means of a so-called gesture control through interactive gestures. With gesture control, the user can select one of the three other control groups from the currently used control group using the gestures "Nod", "Side Tilt to Right" or "Side Tilt to Left". The switch between the four control groups is designed in such a way that the user can decide according to his personal preferences whether to switch using a mouse cursor or gestures. This allows both individual user preferences and any physical limitations to be taken into account when carrying out the interaction gestures. Empirical studies have already shown that older users or users with paraplegia in particular prefer cursor control [19, 20].

Independently of the user's mode of control group switching, the robot is controlled by the position and orientation of the head based on the initial calibrated user-specific rest position. After a quality check and initial calibration, the robot control starts in the idle state, which is displayed graphically on the monitor in a so-called neutral zone. Both the robot and the gripper are stationary, when the neutral zone is active. In addition, the robot stops when the head moves very fast, e.g. when the operator suddenly looks to the side, because something happens there that attracts his attention and distracts him from the task of controlling the robot. This safety mechanism prevents the robot or gripper from moving uncontrolled. Only when the operator starts with the rest position of the head and within the neutral zone, the control is active and the robot or gripper can be manipulated.

This hands-free, head-based control by means of IMU attached to the head is an empirically proven possibility for human-robot interaction. The control is suitable both for workers who need their hands for other tasks and for people with physical disabilities.

4.3   Gesture-Based Control with Myo Armbands

In this control mode the measurement techniques EMG and IMU are used. The Myo armband combines these measuring techniques and integrates them into an armband worn on the upper forearm. Thus, the EMG sensor technology enables the recognition of gestures performed by the hand. Technically, this is made possible by electrodes attached to the bracelet at an equidistant radial distance. Consequently, there is no exact positioning on the muscle, however, the exact sticking of electrodes on the muscle in contrast to the quick attaching of the armband is not a practicable solution for the use in an industrial application. The inaccurate positioning of the electrodes can be compensated by suitable calibration of the armband when it is attached, e.g. by executing an initialization gesture, thus enabling reliable gesture recognition. The IMU allows the relative movement as well as the orientation of the arm in space to be recorded.

The control concept using the Myo armbands envisages a two-handed operation of the robot. By combining the recognized gestures of the left and right hand, the robot switches between control modes within the Myo armband control. Moving the robot with this control is only possible if both hands make the gestures associated with each control mode. As soon as no gesture is detected for a hand, the robot switches to a state in which it does not accept motion commands. This concept ensures maximum safety against unconscious operating malfunctions, as gestures have to be actively performed by both hands for the control unit, and at the same time enables a natural switch-off function of the robot, as the active operation of a switch is no longer necessary. As an additional fallback level, a switch with shutdown function is also available for this operating method. The different control modes are explained below:

- With direct motion control, the robot's Tool Center Point (TCP) follows the relative motion changes measured by the IMU on the operator's right arm. The rotation of the TCP remains unchanged.
- Scaling defines the scale at which the robot follows the relative movements of the operator. A scale of 1:1 is not satisfactory for every application. The operator should equally be able to control movements over the entire movement range of the robot without the operator having to cover the same distance as the robot, as well as the exact positioning of the robot with very small movement changes. The scale is adjusted in this mode by raising or lowering the right arm.
- In TCP rotation control, the rotation changes according to the rotation movements performed by the operator's right arm. Sub-modes allow either changing the rotation with the TCP in a fixed position or changing the rotation with allowed position changes of the TCP by rotation movements.
- The positions of the individual joints or axes of the robot are manipulated by the joint/axis control. For the manipulation of the axes of the linear axis, the relative movements of the arm in the plane are transferred to the respective axis. When controlling the joints of the robot, the respective angle of the joint is increased or decreased by raising or lowering the arm.

Both the direct motion control and the rotation control can be carried out in further sub-modes for specific axes.

4.4 Force-Based Control by Manipulation of TCP

The force-based control follows the idea of guiding the robot by pushing the gripper in the desired direction and is thus an implementation of the kinesthetic teach-in. Sensors measure the forces and torques occurring at the gripper. This makes it possible to determine in which direction the TCP should move or how the rotation of the TCP should be modified. From this information, corresponding control commands for the robot are generated and executed immediately. The robot, therefore, yields to the force and moves in the desired direction. This type of operation represents a very intuitive form of robot control, since no specialist knowledge is required at all. In addition, there is no need to imagine how movements of individual axes or rotations affect the robot's pose, since the robot moves directly to the desired position. In practice, at the work-station CoWorkAs, this type of control can only be used during ongoing operation in the shared workspace of the workstation, in which the human operator can interact with the robot, since the human operator is not safeguarded in the divided workspace. In addition, this control method can also be used outside the running operation for the configuration of movements in the robot workspace. In this case, increased safety requirements must be taken into account and the maximum speed of the robot has to be reduced to a minimum.

4.5   Markerless Motion-Tracking

In this control mode, the technology of depth cameras is used, which makes it possible to capture the position of the human body and its joints. There are two basic variants of how this type of control can be used.

In TCP control, the robot follows the position of a joint or body part of the operator, which is typically the hand. This type of control, for example, can take place in a reference coordinate system, so that the movements of the body parts are transformed into the target coordinate system of the robot. Thus, the robot can be teleoperated without the presence of the operator. Alternatively, the operator's coordinate system can also be selected for control. This approach makes sense when implementing a direct interaction between human and robot and is used, for example, in the adaption of the transfer position mentioned in the workplace description. The closer the robot comes to the human being, the slower it moves. This concept has already been suc-cessfully implemented by Lasota et al. [25]. Shortly before the robot reaches the working person, it stops any movement, so that the human working person has to perform the final approach between the human being and the robot. If the robot detects a contact with the object in the gripper, which is detected by force sensors, the gripper opens, and the transfer of the object is completed. Only if the human moves away from the robot the robot moves again and closes the gripper if necessary.

During joint control of the robot, joint angles are transferred from the human model captured by the depth image cameras to the robot. The joint angles of the Powerball Lightweight Arm LWA 4P used at CoWorkAs are ideally suited for transfer to the joint angles of the human arm, because its structure is very similar to the human arm, especially with respect to the arrangement of the joints. It should be noted that the relative movements of the human hand are usually not transferred to the TCP of the robot with the same position change due to the unequal geometries of the robot and the human arm and the resulting different kinematic chain.

The operation of the robot using a depth camera is advantageous, as markerless tracking is carried out and no additional sensors are required. Thus, this type of control can be used without further effort during the assembly operation.

4.6 Classification of Control Modalities

One goal of the interaction concept was to find at least one control method for each characteristic value of the two dimensions degree of automation and workspace division. Table 1 shows how the presented modalities can be classified according to these dimensions. Computer-based control is suitable for an autonomous operation mode of the robot, since the movements are predefined by the program.

Table 1. Classification approach of control modalities in dimensions degree of automation and workspace division. The filled circles show the degree of general suitability of the control modality for this application are (not filled = not suitable at all, fully filled = highly suitable). For less suitable modalities there is a high implementation effort for a successful realization.

	Shared workspace		Divided workspace	
	autnom.	manual.	autnom.	manual.
Computer-based Control	◐	○	●	○
Head-based Control	○	◐	○	●
Myo gesture Control	○	◐	○	●
Motion-Tracking Control	◐	◐	○	●
Force-based Control	○	●	○	○

Head-based control and Myo gesture control can be aggregated as gesture-based control methods and are therefore only applicable for a manual control mode of the robot. Control by motion tracking is ideal for manual control in teleoperation applications. Computer- and gesture-based control as well as motion tracking methods are mainly suitable for use in divided workspaces. In shared workspaces these control modalities require an additional effort to ensure occupational safety. Additionally, motion tracking control can be used as a technique for an ergonomic interaction design for a mainly autonomous moving robot in a shared workspace, e.g. for position adaptive control.

The forced based control represents a manual control mode, since the robot follows the movements of the operator's hand. Due to the direct interaction between human and robot this interaction method can only be used in shared workspaces.

5 Conclusion and Outlook

By means of the various interaction concepts presented at the HRC research platform CoWorkAs, a broad spectrum of control possibilities for different applications is created. Control modalities are used for the running operation as well as for the configuration of new use cases, which allow for an interaction with the robot even with particular restrictions (e.g. hands-free operation). Special emphasis is placed on the ergonomic, safe and intuitive design of the interaction. At the same time, this framework can be flexibly modified by adding and evaluating further modalities, whereby the implementation effort by using the same basic functionalities of the modular system architecture is kept low.

The next steps are investigations of questions concerning the acceptance, the trust and the usability of the presented control possibilities, since the main purpose of the development of the workplace is the research of interactivity concepts as well as later transfer of the validated concepts into the industry. The knowledge gained is used for the further development and optimization of the concepts.

Acknowledgements. The authors would like to thank the German Research Founda-tion DFG for the kind support within the Cluster of Excellence "Internet of Production (ID 390621612)".

References

1. Masinga, P., Campbell, H., Trimble, J.A.: A framework for human collaborative robots, operations in South African automotive industry. In: IEEM 2015: 2015 IEEE International Conference on Industrial Engineering and Engineering Management: 6–9, December 2015, pp. 1494–1497. Singapore. IEEE, Piscataway, NJ (2015)
2. Audi AG.: Human robot cooperation: KLARA facilitates greater diversity of versions in production at Audi, Ingolstadt (2017)
3. DIN.: EN ISO 10218-2: Robots and robotic devices - safety requirements for industrial robots - Part 2: Robot systems and integration (2012)
4. DIN.: EN ISO 10218-1: Robots and robotic devices - safety requirements for industrial robots - Part 1: Robots (2012)
5. DIN.: ISO/TS 15066 Robots and robotic devices - collaborative robots (2017)
6. Gong, Z., Zhang, Y.: Robot signaling its intentions in human-robot teaming. In: HRI Workshop on Explainable Robotic Systems (2018)
7. Schmidt, L., Herrmann, R., Hegenberg, J., et al.: Evaluation einer 3-D-Gestensteuerung für einen mobilen Serviceroboter. Zeitschrift für Arbeitswissenschaft **68**(3), 129–134 (2014). https://doi.org/10.1007/BF03374438
8. Morais, G.D., Neves, L.C., Masiero, A.A., et al.: Application of Myo armband system to control a robot interface. In: Bahr, A., Abu Saleh, L., Schröder, D., et al. (eds.) Proceedings of the International Joint Conference on Biomedical Engineering Systems and Technologies CMOS Technology, pp. 227–231. Technische Universität Hamburg Universitätsbibliothek; SCITEPRESS - Science and Technology Publications Lda, Hamburg, Setúbal (2016)
9. Sathiyanarayanan, M., Mulling, T., Nazir, B.: Controlling a robot using a wearable device (MYO). In: Int. J. Eng. Dev. Res. (2015)
10. Sanna, A., Lamberti, F., Paravati, G., et al.: A Kinect-based natural interface for quadrotor control. Entertainment Comput. **4**(3), 179–186 (2013). https://doi.org/10.1016/j.entcom.2013.01.001
11. Stowers, J., Hayes, M., Bainbridge-Smith, A.: Altitude control of a quadrotor helicopter using depth map from Microsoft Kinect sensor. In: Gokasan, M. (ed.) IEEE International Conference on Mechatronics (ICM), 2011: 13–15, April 2011, Istanbul, Turkey; proceedings, pp. 358–362. IEEE, Piscataway, NJ (2011)

12. Biao, M., Wensheng, X., Songlin, W.: A robot control system based on gesture recognition using Kinect. Indonesian J. Electr. Eng. Comput. Sci. **11**(5) (2013). https://doi.org/10.11591/telkomnika.v11i5.2493

13. Cheng, L., Sun, Q., Su, H., et al.: Design and implementation of human-robot interactive demonstration system based on Kinect. In: 24th Chinese Control and Decision Conference (CCDC), 2012: 23–25, May 2012, pp. 971–975. Taiyuan, China. IEEE, Piscataway, NJ (2012)

14. Du, G., Zhang, P., Mai, J., et al.: Markerless Kinect-based hand tracking for robot teleoperation. Int. J. Adv. Rob. Syst. **9**(2), 36 (2012). https://doi.org/10.5772/50093

15. Song, W., Guo, X., Jiang, F., et al.: Teleoperation humanoid robot control system based on Kinect sensor. In: 4th International Conference on Intelligent Human-Machine Systems and Cybernetics (IHMSC), 2012: 26–27, Aug. 2012, pp. 264–267. Nanchang, Jiangxi, China. IEEE, Piscataway, NJ (2012)

16. Li, C., Yang, C., Wan, J., et al.: Teleoperation control of Baxter robot using Kalman filter-based sensor fusion. Syst. Sci. Control Eng. **5**(1), 156–167 (2017). https://doi.org/10.1080/21642583.2017.1300109

17. Rudigkeit, N., Gebhard, M., Gräser, A.: Towards a user-friendly AHRS-based human-machine interface for a semi-autonomous robot. In: 2014 IEEE/RSJ International Conference on Intelligent Robots and Systems, Workshop on Assistive Robotics for Individuals with Disabilities: HRI Issues and Beyond (2014)

18. Jackowski, A., Gebhard, M., Graser, A.: A novel head gesture based interface for hands-free control of a robot. In: 2016 IEEE International Symposium on Medical Measurements and Applications, pp. 1–6. IEEE, Piscataway, NJ (2016)

19. Nelles, J., Kohns, S., Spies, J., et al.: Analysis of stress and strain in head based control of cooperative robots through tetraplegics. World Acad. Sci. Eng. Technol. Int. J. Med. Health, Biomed. Bioeng. Pharm. Eng. **11**(1), 11–22

20. Nelles, J., Schmitz-Buhl, F., Spies, J., et al.: Altersdifferenzierte Evaluierung von Belastung und Beanspruchung bei der kopfbasierten Steuerung eines kooperierenden Roboters. In: Frühjahrskongress der Gesellschaft für Arbeitswissenschaft (2017)

21. Caccavale, R., Saveriano, M., Finzi, A., et al.: Kinesthetic teaching and attentional supervision of structured tasks in human–robot interaction. Auton. Robots **57**(5), 469 (2018). https://doi.org/10.1007/s10514-018-9706-9

22. Ruffaldi, E., Di Fava, A., Loconsole, C., et al.: Vibrotactile feedback for aiding robot kinesthetic teaching of manipulation tasks. In: Human-robot collaboration and human assistance for an improved quality of life: IEEE RO-MAN 2017: 26th IEEE International Symposium on Robot and Human Interactive Communication, August 28-September 1, 2017, pp. 818–823. Lisbon, Portugal. IEEE, Piscataway, NJ (2017)

23. Kormushev, P., Calinon, S., Caldwell, D.G.: Imitation learning of positional and force skills demonstrated via kinesthetic teaching and haptic input. Adv. Robot. **25**(5), 581–603 (2011). https://doi.org/10.1163/016918611X558261

24. Gammieri, L., Schumann, M., Pelliccia, L., et al.: Coupling of a redundant manipulator with a virtual reality environment to enhance human-robot cooperation. Procedia CIRP **62**, 618–623 (2017). https://doi.org/10.1016/j.procir.2016.06.056

25. Lasota, P.A., Rossano, G.F., Shah, J.A.: Toward safe close-proximity human-robot interaction with standard industrial robots. In: IEEE International Conference on Automation Science and Engineering (CASE), 2014: 18–22, Aug. 2014, pp. 339–344. Taipei, Taiwan. IEEE, Piscataway, NJ (2014)

Determination of the Subjective Strain Experiences During Assembly Activities Using the Exoskeleton "Chairless Chair"

Sandra Groos[✉], Marie Fuchs, and Karsten Kluth

Department Mechanical Engineering, Ergonomics Division,
University of Siegen, Paul-Bonatz-Straße 9-11, 57068 Siegen, Germany
{groos,kluth}@ergonomie.uni-siegen.de,
marie.fuchs@student.uni-siegen.de

Abstract. The passive exoskeleton "Chairless Chair" has been studied in the past for its practical use, but the true physiological benefits have not yet been thoroughly researched. Therefore, 17 subjects carried out, with and without the exoskeleton, three series of tests in the laboratory (treadmill ergometer, screwing and assembly tasks) in order to expand the knowledge about the benefits, but also about the limitations of the exoskeleton. Besides an objective stress and strain measurement, the physiological analysis was completed by the systematic interview of subjective strain perception. The subjective results show that the use of the exoskeleton during screwing and assembly work was perceived as beneficial. Walking on the treadmill, on the other hand, was viewed much more critically. The exoskeleton received negative ratings, especially with regard to posture and the feeling of safety. Many test persons therefore saw potential for improvement in the design as well as in some safety aspects.

Keywords: Industrial exoskeletons · Assembly workstations · Leg support · Back support · Unfavorable postures · Subjective strain · Systematic interview

1 Introduction

Above all, demographic change, the resulting ageing workforce and the lack of qualified junior staff are the positive reasons why companies are increasingly being forced to act when it comes to workplace design. The main objective is to ensure that the work can be carried out permanently and without short or long-term physical or psychological impairments. Ergonomic workplace design is one of many measures that contribute to the employee being able to work at his or her workplace until retirement and thus actively counteract the growing shortage of skilled workers.

Nevertheless, there are still a large number of jobs where workers have to endure physically heavy work under additional restraints. Often, however, the possibilities of ergonomic workplace design are simply exhausted, or an adaptation of the workplace to the employee is not possible. In the industry of heavy and special machine construction, for example, it is often difficult to adjust the working area to a healthy working height due to the dimensions and weights of the components to be processed. In such cases, different types of exoskeletons can help. An exoskeleton is a mechanical

© Springer Nature Switzerland AG 2020
J. Chen (Ed.): AHFE 2019, AISC 962, pp. 72–82, 2020.
https://doi.org/10.1007/978-3-030-20467-9_7

structure designed to support the user muscularly [1, 2]. Based on the first developments of the so-called "Hardiman" [3], exoskeletons are now largely established in the military and medical sectors. But exoskeletons are not yet very widespread at commercial workplaces [4], especially in Europe, even though their suitability for practical suitability is tested by prototypes in industrial companies (primarily automobile manufacturers) [5, 6].

During the development process of exoskeletons, but also in the final implementation into operational practice, acceptance on the part of the user must be as important as the physiological benefit. This is the only way to ensure successful operational implementation [7]. But there is still an increased perception of complaints in the human/exoskeleton interface areas, which must be counteracted, even though the use of an exoskeleton is regarded as supportive and therefore helpful [8]. In addition to user acceptance, the potential hazards posed by exoskeletons (e.g. an increased risk of tripping and falling accidents) are just as important, but they are still largely unexplored [4].

In order to obtain initial insights into the stress and strain associated with the use of the passive exoskeleton "Chairless Chair" from the Swiss company noonee AG, a laboratory study was carried out in which test persons executed three different tasks.

2   Test Design and Methodology

The test design and methodology is described through the test procedure, the test subjects and the analysis of the subjective strain experiences in detail in the following sub-chapters.

2.1   Test Procedure

For the analysis of the stress and strain during the use of a passive exoskeleton for leg and back support, 17 subjects (Ss), inexperienced in using exoskeletons, carried out simulated work situations. The work simulation was subdivided into partial tests and consisted of a treadmill analysis, a screwing test and an assembly work, whereby all tests were carried out with and without exoskeleton. In order to prevent certain habituation and fatigue effects, the partial tests were randomized.

Table 1 shows the test procedure of one day as an example of one subject. In addition to the general briefing and information about the contents and risks of the tests and the obtaining of the declaration of consent, the application of the measuring technique for objective data acquisition of heart rate and energy expenditure took place. Subsequently, the body-relevant data were asked for and measured, and existing muscular tensions and pain in the musculoskeletal system were systematically questioned. Before the actual partial tests, a detailed instruction in the use of the "Chairless Chair" took place, in which in particular the safe use of the exoskeleton was practiced.

Before the respective partial tests (consisting of treadmill analysis, screwing and assembly work) were carried out, the Ss were instructed in the specific work tasks. After each of the partial tests described below, the subjective stress experience was evaluated systematically. The test day ended with the removal of the measuring equipment and a final survey.

Table 1. Test schedule for one subject.

Length [min]	Task
40	General briefing and instruction, application of the measuring technique, explanation of the questionnaire and collection of the body-relevant data, collection of resting pulse and metabolic rate at rest, instruction in handling the exoskeleton
2	Introduction in the screwing test
15	Screwing test without exoskeleton (PE3), followed by questionnaire
2	Instruction in the assembly work test and installation of the exoskeleton
20	Assembly work test with exoskeleton (PE6), followed by questionnaire
15	Screwing test with exoskeleton (PE4), followed by questionnaire
2	Introduction to treadmill analysis
15	Treadmill analysis with exoskeleton (PE2), followed by questionnaire and removing the exoskeleton
20	Assembly work test without exoskeleton (PE5), followed by questionnaire
15	Treadmill analysis without exoskeleton (PE1), followed by questionnaire
10	Removal of the measuring technique
15	Final survey
	End

Partial Experiment 1 and 2 – Treadmill analysis. For the treadmill analysis (PE1 and PE2), the Ss walked with and without exoskeleton 10 min on a treadmill, each. The speed was 2.4 km/h (approx. 1.5 mph) with an incline of 0% (see Fig. 1). A higher speed could not be chosen in terms of occupational safety and hazard avoidance, in particular due to the use of the exoskeleton. No further tasks had to be performed during walking.

Fig. 1. Treadmill analysis without (on the left) and with exoskeleton (on the right).

Partial Experiment 3 and 4 – Screwing test. The screwing test consisted of a 5-min screwing with a hand screwdriver and a cordless screwdriver, each, into a medium-density wood fiber board with prefabricated drill holes of different heights (cf. Fig. 2). Since the aim of the experiment was to achieve the most unfavorable posture possible, the working height was set to 50–55 cm (1'8"–1'10"), independent of the test person. In order to be able to make statements about the work productivity the number of screwed in screws, but also the dropped screws, was noted.

Fig. 2. Screwing test with (on the left) and without exoskeleton (on the right), using a hand screwdriver and a cordless screwdriver, each.

Partial Experiment 5 and 6 – Assembly work. In the assembly work test, wooden and plastic components were assembled and dismantled according to given photo models. The working speed was chosen by the test person, so that the process had to be repeated continuously over a period of 15 min. The working height was again set as unfavorably as possible at 45 cm (1'6") (cf. Fig. 3).

Fig. 3. Assembly work activity with (on the left) and without exoskeleton (on the right).

2.2   Test Subjects

The choice of the test persons (Ss) was only limited with regard to body size, as the manufacturer stated that the "Chairless Chair" should only be used with a body height between 1.60 m and 1.90 m (5'3"–6'3"). However, leg length and hip height have proved to be much more decisive in assessing safe use. The body-relevant data of the Ss that participated in the test are compiled in Table 2. The four-stage adjustment of the length of the upper and lower leg support of the "Chairless Chair" (XS-L) was carried out by trained personnel on the basis of the body measurements.

A total of 17 persons (10 male, 7 female) who had no experience at all with the "Chairless Chair" took part in the experiments. The average age was 26.7 years, which meant that most of the participants were younger.

Table 2. Specific information on the 17 test subjects (Ss)

Subject ID	Age [Years]	Weight [kg]	Size [cm]	Waist height [cm]	Gender [m/f]	Setting exoskeleton [upper leg/lower leg]
01	34	66	158	86	f	XS/XS
02	39	96	175	97	m	S/M
03	40	89	185	110	m	L/L
04	23	70	173	101	f	S/S
05	23	82	173	93	m	M/M
06	21	66	178	103	m	M/M
07	22	78	185	109	m	M/M
08	25	75	177	104	m	M/M
09	26	92	172	98	m	S/M
10	24	76	176	108	f	M/M
11	28	85	176	107	f	M/M
12	19	75	185	112	m	L/L
13	24	57	175	102	f	M/M
14	25	83	173	99	f	M/S
15	24	70	185	108	f	L/L
16	28	78	189	106	m	L/L
17	29	75	188	112	m	L/L
$\bar{x} \pm S_d$	26.7 ±5.9	77.2 ±10.0	177.8 ±7.7	103.2 ±7.0		

2.3   Analysis of the Subjective Strain Experiences

The subjective perception of the test persons was recorded before, during and after the experiments by means of an interview. Before the experiments were carried out, a body diagram was used to record the current physical constitution. For this purpose, muscular tensions, muscle soreness of injuries and their subjectively perceived intensity were systematically recorded on a scale from 0 (no impairment) to +4 (very severe impairment). After each of the six partial experiments, an individual evaluation of the subjectively felt physical strain was carried out again. The body diagram and the scaling were identical to those before the experiment, but the type of impairment was differentiated into muscular tension, pressure points and numbness.

For the comparative evaluation of the tests with and without exoskeleton, a survey was also carried out using closed questions in simple checkbox questionnaires. Both bipolar 4-stage scales and monopolar 4-stage scales were used, which were also visually supported by the Kunin face scale [9]. The most important questions from the subjective survey can be taken from the following results.

3 Results

The results gained from the subjective evaluation of the experienced strain are presented below, subdivided into the three different activities.

3.1 Partial Experiment 1 and 2 – Treadmill Analysis

During treadmill activity without an exoskeleton, none of the Ss suffered any impairment at any of the body parts interrogated. When walking with the exoskeleton, however, slight impairments (mean values between 0.1 and 0.3 in a 4-step scale) were observed. Muscular tensions mainly affected the shoulder and neck area as well as the thighs. Pressure points occurred exclusively in the human/exoskeleton contact points, i.e. on the hip, thighs, ankles and feet. While the impairment occurring with the exoskeleton on the treadmill are rather negligible due to their low intensity, further partial aspects of the subjective questioning were regarded as far more critical. As shown in Fig. 4, the evaluation of the posture during the two partial experiments with and without exoskeleton showed values of $+3.6 \pm 1.0$ (without exoskeleton) and $+1.6 \pm 2.2$ (with exoskeleton) at a bipolar scale of -4 (very unfavorable) to $+4$ (very favorable).

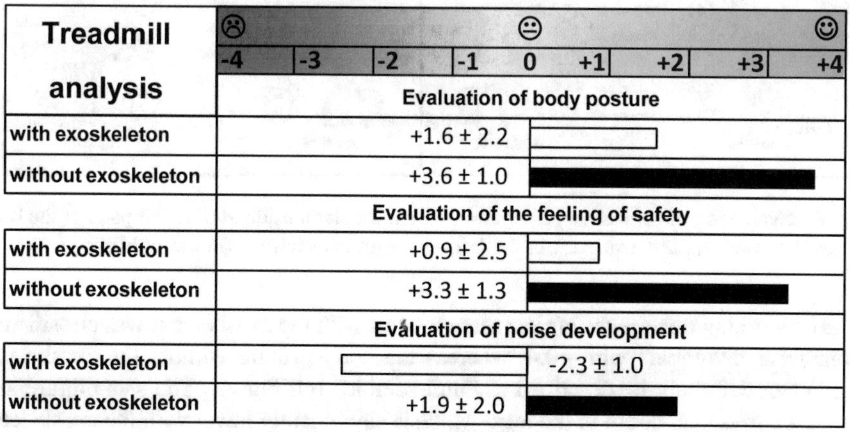

Treadmill analysis	☹ -4	-3	-2	-1	☺ 0	+1	+2	+3	☺ +4
	Evaluation of body posture								
with exoskeleton	$+1.6 \pm 2.2$								
without exoskeleton	$+3.6 \pm 1.0$								
	Evaluation of the feeling of safety								
with exoskeleton	$+0.9 \pm 2.5$								
without exoskeleton	$+3.3 \pm 1.3$								
	Evaluation of noise development								
with exoskeleton	-2.3 ± 1.0								
without exoskeleton	$+1.9 \pm 2.0$								

Fig. 4. Evaluation of subjective perception of body posture, feeling of safety and noise development during treadmill analysis with and without exoskeleton.

The difference was even more pronounced when assessing the feeling of safety on the treadmill. Without exoskeleton the Ss felt significantly safer (+3.3 ± 1.3) than with exoskeleton (+0.9 ± 2.5). The development of noise when walking with the exoskeleton was perceived as particularly disturbing (−2.3 ± 1.0), whereas walking without the exoskeleton was perceived as pleasant with +1.9 ± 2.0. Finally, all 17 Ss stated that they could walk better without the exoskeleton than with the exoskeleton.

3.2   Partial Experiment 3 and 4 – Screwing Test

When performing the screwing test (with a hand and a cordless screwdriver) without the support of the exoskeleton, mainly slight muscular tensions occurred in almost all body areas (0.1–0.8 on the 4-stage scale). Only in the lumbar vertebrae were the tensions slightly more pronounced with 1.2 (cf. Fig. 5). Due purely to the activity, slight pressure points in the right hand could also be recorded. The use of the exoskeleton could reduce the tension only in the lumbar vertebra area from originally 1.2 to 0.4 during the performance of the activity. In all other areas there were hardly any improvements due to the changed body posture, rather pressure points were again found in the hip, thigh and lower leg areas.

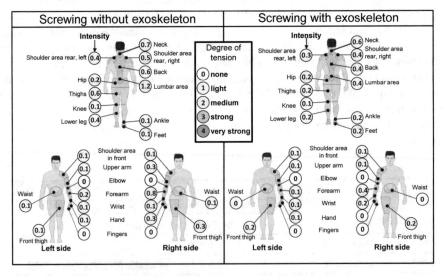

Fig. 5. Evaluation of subjective perception of the muscular tension at different parts of the body during the screwing test without (on the left) and with exoskeleton (in the right).

The working posture during the screwing test with the exoskeleton was classified as "somewhat favorable" with +0.4, whereas the body posture without the exoskeleton (−2.1) can definitely be described as "unfavorable" (cf. Fig. 6). The same figure also shows the muscular strain in the legs, which is significantly lower with the exoskeleton (−1.1). The evaluation of the feeling of safety is rated significantly higher for the use of the exoskeleton in this test than during the treadmill analysis, but the Ss still felt

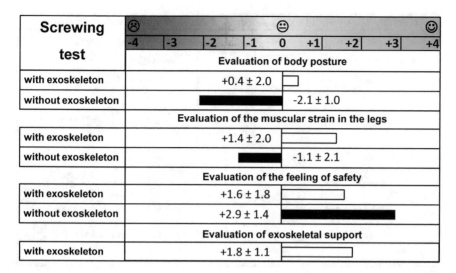

Screwing test	☹				☺				☺
	-4	-3	-2	-1	0	+1	+2	+3	+4
	Evaluation of body posture								
with exoskeleton				+0.4 ± 2.0					
without exoskeleton			-2.1 ± 1.0						
	Evaluation of the muscular strain in the legs								
with exoskeleton				+1.4 ± 2.0					
without exoskeleton			-1.1 ± 2.1						
	Evaluation of the feeling of safety								
with exoskeleton				+1.6 ± 1.8					
without exoskeleton			+2.9 ± 1.4						
	Evaluation of exoskeletal support								
with exoskeleton				+1.8 ± 1.1					

Fig. 6. Evaluation of subjective perception of body posture, muscular strain in the legs, feeling of safety and the overall exoskeletal support during the screwing test (with hand and cordless screwdriver) with and without exoskeleton.

significantly less safe (+1.6) than for screwing without the exoskeleton (+2.9). Overall, however, the Ss rated the exoskeleton support as quite good with +1.8. Thirteen of the 17 interviewed persons would choose the "Chairless Chair" for support when working with screws at this working height.

3.3 Figures Partial Experiment 5 and 6 – Assembly Work

During the simulated assembly work and without the support of the "Chairless Chair", medium to severe tensions occurred primarily in the lumbar vertebrae area and in the thighs. Figure 7 illustrates the degree of tension in the body parts evaluated during assembly work without (left) and with (right) exoskeleton. The figure also shows that the support provided by the exoskeleton in the critical areas results in a significant reduction of tension. Only in the shoulder and neck area do the muscular tensions increase. In this partial experiment, slight pressure points also occurred at the human/exoskeleton interfaces.

Overall, the working posture could be significantly improved by using the "Chairless Chair". Starting from an average value of −3.6, the value was improved to −0.7, but was nevertheless perceived as somewhat unfavorable (cf. Fig. 8). Muscular stress in the legs was also reduced from −2.7 to +0.1. However, the feeling of safety during this simulated assembly work was also slightly lower when using the exoskeleton (+1.0) than without the exoskeleton (+1.9). Nevertheless, exoskeletal support was rated as good overall (+2.2). Fifteen of the 17 interviewed persons would decide to use the "Chairless Chair" for such a type of work.

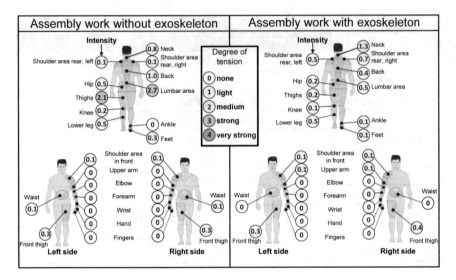

Fig. 7. Evaluation of subjective perception of the muscular tension at different parts of the body during the assembly work without (on the left) and with exoskeleton (in the right).

Assembly	☹				☺				☺
	-4	-3	-2	-1	0	+1	+2	+3	+4
work				Evaluation of body posture					
with exoskeleton					-0.7 ± 2.0				
without exoskeleton					-3.6 ± 0.7				
			Evaluation of the muscular strain in the legs						
with exoskeleton				+0.1 ± 1.7					
without exoskeleton					-2.7 ± 1.7				
				Evaluation of the feeling of safety					
with exoskeleton				+1.0 ± 2.1					
without exoskeleton				+1.9 ± 2.1					
				Evaluation of exoskeletal support					
with exoskeleton				+2.2 ± 1.1					

Fig. 8. Evaluation of subjective perception of body posture, muscular strain in the legs, feeling of safety and the overall exoskeletal support during the simulated assembly work with and without exoskeleton.

4 Discussion

During the treadmill analysis, the test persons' low feeling of safety turned out to be the biggest problem. On the one hand, the user is required to increase his concentration in order to avoid unwanted accidents, and on the other hand, it is difficult to compensate accidental movements such as loss of balance. Even if the "Chairless Chair" is not designed to cover longer distances, the main problem is that it is impossible to climb stairs with the exoskeleton.

The particularly disturbing noise development is essentially due to two design features. The joint between the upper and lower leg support can only be rotated up to an angle of about 180°. When walking, both leg supports move with each other so that an audible stop occurs in the joint with every step forward (extension to 180°). Furthermore, due to the anatomical characteristics of some Ss, it was not possible to place the rubber socket of the lower leg support a few centimeters above the floor in a standing posture, as actually intended by the manufacturer. As a result, the rubber nozzle was placed on the floor with every step and caused an additional noise. However, the new food module of the "Chairless Chair", which could not yet be considered in this study, can now better counteract this problem.

During the screwdriving tasks with the hand and cordless screwdriver, the support provided by the "Chairless Chair" was subjectively rated very positively. The same applies to the simulated assembly activity in which the unfavorable posture and the associated static holding work of the large muscle groups could be significantly reduced. Nevertheless, the Ss did not feel as safe as hoped with the support of the exoskeleton during these experiments. However, these circumstances can also be attributed to a lack of experience in dealing with the exoskeleton.

5 Conclusion

The present laboratory study was able to demonstrate the physiological benefit on the basis of a subjective evaluation during the screwing test and assembly work. However, the subjective evaluation of the Ss also revealed some design weaknesses. The foot module of the "Chairless Chair" has already been improved so that the exoskeleton can be put on and taken off much more easily and the safety in a seated position can be increased. The latter is due in particular to the fact that the new foot module enables the foot to be pushed in front of the knee in a sitting position, which ultimately increases stability. Trained specialists should adjust the size of the leg supports since the anatomical characteristics of the human being, i.e. different leg lengths for the same body size, only allow a limited adjustment based on the body size (as indicated on the product). In order to increase user acceptance, further technical development should also focus on the avoidance of pressure points, which frequently occur at the human/exoskeleton interface.

The results, which are clear despite the short duration of the tests, suggest that a prolongation of the tests produces even clearer results, which is why efforts should be made to repeat the series of tests with an extended and thus realistic performance of the work. Likewise, in order to reflect the population of persons employed in production better, the test persons' collective should me more heterogeneous with regard to age.

Overall, however, it should be noted that the use of the "Chairless Chair" showed a clear physiological benefit and was regarded by the Ss as supportive while performing the work tasks. Nevertheless, the long-term effects as well as the question of the daily maximum usage time of the exoskeleton must also be researched in further studies.

References

1. Herr, H.: Exoskeletons and orthoses: classification, design challenges and future directions. J. NeuroEng. Rehabil. **6**, 21 (2009)
2. de Looze, M.P., Bosch, T., Krause, F., Stadler, K.S., O'Sullivan, L.W.: Exoskeletons for industrial application and their potential effects on physical work load. Ergonomics **59**(5), 671–681 (2016)
3. General Electric: Research and development for machine augmentation of human strength and endurance – Hardiman I Project. New York (1971)
4. Schick, R.: Einsatz von Exoskeletten an gewerblichen Arbeitsplätzen. DGUV Forum **1-2**, 8–11 (2018) German only
5. Hensel, R., Keil, M.: Subjektive Evaluation industrieller Exoskelette im Rahmen von Feldstudien an ausgewählten Arbeitsplätzen. Z. Arb. Wiss. **72**, 252—263 (2018) German only
6. Spada, S., Ghibaudo, L., Carnazzo, C., Di Pardo, M., Chander, D.S., Gastaldi, L., Cavatorta, M.P.: Physical and virtual assessment of a passive exoskeleton. In: Bagnara, S., Tartaglia, R., Albolino, S., Alexander, Th., Fujita, S. (eds.) Proceedings of the 20th Congress of the International Ergonomics Association (IEA 2018). Volume VIII: Ergonomics and Human Factors in Manufacturing, Agriculture, Building and Construction, Sustainable Development and Mining, pp. 247–257. Springer Cham (2019)
7. Steinhilber, B., Seibt, R., Luger, T.: Einsatz von Exoskeletten im beruflichen Kontext – Wirkung und Nebenwirkung. Arbeitsmedizin – Sozialmedizin – Umweltmedizin **53**, 662–664 (2018) German only
8. Bosch, T., van Eck, J., Knitel, K., de Looze, M.P.: The effects of a passive exoskeleton on muscle activity, discomfort and endurance time in forward bending work. Appl. Ergon. **54**, 212–217 (2016)
9. Kunin, T.: The construction of a new type of attitude measure. Person. Psych. **8**, 65–78 (1955)

A Nursing Robot for Social Interactions and Health Assessment

Anja Richert[✉], Michael Schiffmann, and Chunrong Yuan

Cologne Cobots Lab, TH Köln - University of Applied Sciences,
Betzdorfer Str. 2, 50679 Cologne, Germany
{anja.richert, michael.schiffmann,
chunrong.yuan}@th-koeln.de

Abstract. Social Robotics for nursing care gain increasing importance in the age of demographic change and fast development of artificial intelligence. Social robots are automated or autonomous machines capable of interacting with people on the basis of social rules and are mostly humanoid and mobile. The Cologne Cobots Lab (Cologne Cobots Lab is an interdisciplinary research lab of the TH Köln – University of Applied Sciences, with its main research focus in the areas of collaborative and social robotics) is currently carrying out research in social and cognitive robotics. In the context of nursing care, approaches for capturing emotions and the state of mind of patients through different interaction analytics will be developed. We will use the humanoid robot pepper and extend its software functions so that it can take initiatives in human-machine-conversation. We conduct different user-centered experiments in real-world conditions and investigate the verbal interactions among human users and the robot system. In this regard, it would be both valuable and interesting to find out whether an AI enabled humanoid robot is able to stimulate or encourage conversation or even perform better than a natural conversation partner.

Keywords: Nursing care · Human robot collaboration · Social interaction · Natural language analytics · Health assessment · Social robotics

1 The Polylemma in Nursing Care

The nursing care sector in Germany is facing several dilemmas with multiple factors interacting with each other. This leads essentially to a polylemma:

a) The population in Germany is subject to a rapid ageing process, i.e. the number of people in need of long-term care is continuously increasing. Assuming that age- and gender-specific care rates remain unchanged and that the population develops according to the 13th coordinated population projection, the number of people in need of care would rise to 4.7 million by 2060 [1].

b) A shortage of personnel and time for nursing staff is a major problem, especially in-patient care. At the same time, older nursing staff is leaving the profession early due to the high physical and psychological strain.

c) The nursing sector has an above-average sickness rate. An analysis of the inca-pacity to work reported by the Scientific Institute of the AOK (WIdO), a medical

© Springer Nature Switzerland AG 2020
J. Chen (Ed.): AHFE 2019, AISC 962, pp. 83–91, 2020.
https://doi.org/10.1007/978-3-030-20467-9_8

insurance company, showed that 6.3% of employees in patient nursing care of the elderly are absent due to illness every day, while the national average sickness rate in all sectors is only 4.8%. In addition, long-term illnesses dominate in the care sector, which means that employees are often unable to work for more than four weeks [2].

In order to solve the above mentioned polylemma, systemic changes have to be made in the area of nursing care in Germany. At present, one can observe a kind of "catching-up modernization" in the sector, where the strategy lies mainly in the implementation of new technology (cf. [3]). During the last decade, assistive systems based on ICT (Information and Communication Technology) have been the main focus of research. Examples of such kinds of systems are electronic patient records, digital assistive and monitoring systems. Recently a technologically shift has taken place, where the use of robotic systems plays a central role and has become the new focus of research in nursing care. In the near future, properly designed assistance systems and care robots could be made available for providing the necessary care and services in a natural environment for people who may need them. This will extend the time period in which people live independently. At the same time, such kinds of robot systems could also collect medical data and if necessary also communicate with their counterparts (cf. ibid.). As a result, care personnel and patients can spend more quality time with each other and the overall quality of caring services can be improved. Quality in care is described in this context as a dynamic, reciprocal relationship between the system, interventions, clients and outcome, based on the Quality Health Outcome Model [4]. This is where the SOHO (State of Health Robotics) project comes in. In this project, we aim to achieve a robot-supported health assessment which is built upon assistive elements and dialogue functions. These elements and functions are set up based on our discussions with care specialists. We use the well-established robotic platform Pepper for staying with the patients, where it should act as an initiative conversation partner, provide assistant functions (e.g. remind them of taking medication, call attention of their drinking behavior, answering questions etc.) and analyze the health conditions reported orally by the patients themselves (Patient Reported Outcome Measurement - PROMs).

To this end, SOHO not only uses standardized analogue methods (PROMs) and make them accessible, but also makes further development of specific methods (e.g. UCLA Loneliness Scale, cf. [5]) within the framework of digital, robotic technology. In addition, a novel approach, i.e., state measurement via NLP (Natural Language Processing), will be developed and tested. In order to perform this kind of experiments in clinic and home environment, we will develop software components to be integrated in Pepper for improving its cognitive/interaction functions/capabilities. The final system is supposed be able to initiate a conversation once a person is perceived, remind patients on drinking enough water or taking medications, answering questions and behave according to the mood of its interlocutor. During the daily caring process, the robot should collect information about the physical and emotional status of patients and transform it into a "health assessment summary". The pre-registered caregivers and doctors should get access to the summary and are thereby be supported with a higher level of patient information so as to provide the exact caring process suitable for individual patients. For the purpose of quantification, we not only use field tested psychological scales, but also develop further context and domain specific scales, aiming at finding out the measurements which can characterize the underlying disease patterns. With such kind of measurements and indications, it would be possible to do a fast detection and prediction of certain diseases and the patients could have better chance of getting an earlier treatment and much more to track known diseases of a patient and their clinical course.

SOHO starts with a conceptual analysis and specification of the necessary functions for the robot system. A human centered design procedure is adopted for ensuring an accurate requirement engineering and test procedures together with the patients and caregivers. Once the necessary functions have been fully integrated on the Pepper robot, the experimental study will begin and the experimental data will be collected, analyzed and saved. Taking the entire product life into account, we will also analyze the financing, maintenance, performance and cost efficiency of such a system, so as to suggest sound business models for enabling robotic use in nursing care.

2 State of the Art Robots in Nursing Care

Physical support is only one way to implement robot technologies in care. The acceptance of such assistance technologies, which work in close cooperation with humans, is currently rather low, at least in Europe. Based on a systematic analysis of the state of nursing robotics three categories according to their functionalities and areas of applications can be differentiated: Socio-assistive systems (including emotion robotics), service robotics and rehabilitation robotics [3]. In the context of our work, the areas of socio-assistive systems and service robotics are of particular importance. Therefore, we focus on these two categories in our analysis of the state-of-the-art. In parallel, the improvements of our approach will be elaborated and discussed.

2.1 Socio-Assistive Systems

Socio-assistive systems are robotic solutions that focus on the social-communicative aspects of care. Bemelmans et al. [6] undertakes a comprehensive secondary analysis of the international research studies on the socio-assistive systems developed at that time for supporting older people. Through this analysis, it is shown that such systems have predominantly positive effects. However, due to the limited coverage and quality of the individual studies, the results are not very significant [cf. 3, 6]. Socio-assistive systems in the form of humanoid robots have also been investigated, considering different aspects including usability and linguistic interaction. During the experimental study, the test persons, including older people who had no previous contact with computers, expressed their interest and satisfaction with the systems used [7]. In the field of care, the seal-robot "Paro" has received comparatively high attention. The system is intended to promote relaxation and improve the overall social and psychological situation of patients [8]. Even though "Paro" is currently one of the most researched systems in the field of nursing robotics, the positive findings achieved in the various studies cannot be described as representative in any way [9]. Softbank Robotics (formerly Aldebaran) has launched several successful robotic systems such as "Nao" and "Pepper". Both are designed as an adaptive robot companion that can not only move, speak and dance but also interact with humans on an emotional level. We have applied AI techniques and developed different software modules for the Nao robot so as to extend its cognitive capabilities and achieve better audio and visual interactions [cf. 10]. As a successor of Nao, Pepper has already been used in the healthcare sector. In the project ARiA funded by the Federal Ministry of Education and Research, Pepper is used as a prototype in an old people's home to pass the time and activate senior citizens. Applications such as memory games, singing and dancing are already being tested there.

Socio-assistive systems have so far tended to be characterized by long development periods. In the case of Paro, for example, it took 11 years before it was ready for series production. While the production of speech, facial expressions, gestures, etc. is partly feasible, the recognition of human expressions, which is essential for successful social interaction, is still afflicted with serious problems and needs improvement [cf. 11]. The development of an autonomous social robot with complex social behavior goes beyond the limits of conventional robotics research [cf. 12]. In addition to the original robotic topics such as learning, adaptation, planning, control technology, etc., new topics such as automatic adaptation to different human behavior patterns, attention control or the recognition of emotions and intentions are added [12].

2.2   Service Robotics

Research and development often focus on robotic systems to provide simple services, as the complexity is lower compared to socio-assistive systems [3]. In the context of a meta-analysis of autonomous systems to support elderly and very old people, the current developments are often technology-driven ("technology-push", see also [13]) and it is urgently recommended that the needs of elderly people should be given much greater consideration in the future [14]. Autonomous systems to support older people are therefore still insufficiently mature and cannot yet make a substantial contribution to an independent life (cf. also [3]). The majority of service robots currently in research and development have still very limited capabilities in obstacle avoidance and provide insufficient manipulation possibilities. Thus, they are more suitable for robot-oriented, i.e. clearly structured, than for human-oriented, i.e. unstructured, working conditions. Previous prototypes such as "PR2" (Willow Garage), "Twendy-One" (Waseda University), "Care-O-Bot" (Fraunhofer IPA), "Rollin' Justin61" (DLR), "Armar" (Karlsruhe Institute of Technology) or "Cosero" (University of Bonn) are designed rather as research equipment than as a commercial product for competitive use in practice. A productive use of such systems under real conditions is far from conceivable. One of the most advanced systems, which is already very close to market maturity, is the Care-O-Bot developed by the Fraunhofer Institute for Manufacturing Engineering and Automation, which has been available in its fourth generation since 2015. The robot, which thanks to innovative ball joints in the hip and neck, can even bend and master simple gestures such as pitching or shaking the head. It has a modular design and can therefore be configured for a wide range of applications. The originally intended application purpose of the Care-O-Bot, as the name already suggests, is for nursing use at home or in a professional facility. A typical application scenario of it, which has been tested in real life, is fetching or bringing objects so that the caregivers can concentrate on their actual task. The system is operated via a touch screen attached to the head and additionally has microphones for speech recognition and cameras for person and gesture recognition. However, the system is currently "only" used in by a store chain of electronics (Saturn) and a museum (Haus der Geschichte, Bonn).

The current publications and discussions on the subject show substantial increase of research and development activities of nursing robotics but render only limited information regarding the conditions of their use in real-world and how well the caring tasks can be achieved. Caring robots appear mainly as socio-assistive systems or service robotics. Socio-assistive systems for supporting social-communicative aspects of care favor mostly humanoid systems or platforms with animal shapes. The reason may be associated with the facts that such systems are able to attract the attention and interest of aid recipients and hence lead to improved interaction behavior. The available studies are still methodologically inadequate and usually focus fragmentarily on the use of a few selected systems. Service robotics, which are currently developed for the area of care, consider mostly such aspects like mobility, self-care, interaction or relationship with aid recipients. Service robotics, which are developed for the use by professional caregivers, address primarily the logistic and organizational aspects of care work. There exist hardly any systems which could support direct and patient-oriented care work.

The complexity of the caring services is increased by the fact that the mindset of restricting care work exclusively to the relationship caregiver-recipient is nowadays no longer viable [cf. 3]. In future, nursing work will be provided by hybrid teams (human-robot teams) as a "cooperative service", where the exact conditions for possible success have to be extensively researched.

3 Social Interactions and Health Assessment with Robot

3.1 Main Concept

The overall objective of the SOHO project is to develop a social robotic system that supports nursing care and improves the quality of care. For this purpose, the robotic system should assume communicative functions such as serving as an initiating conversation partner, assistance functions (reminding patients to take medication, drinking behavior, etc.) and analyses the patient's subjective state of health (Patient Reported Outcome Measurement - PROMs). By outsourcing the schematic and recurring communication components (such as tablet intake queries, etc.) to robotic systems, the nurse has more time for more in-depth discussions with the patients. In addition, patient related data acquisition may become more accurate due to the permanent presence of the robotic system, as data is queried at the time of its formation and not retrospectively in cognitively impaired patients. In addition, the thesis must be investigated that the presence of a robotic system has a positive effect on the emotional well-being of the patients, for example by generating a cheerful mood through the communicative functions, by preventing loneliness, etc. The presence of a robotic system can also be used as a means of preventing the emotional state of the patients. SOHO therefore takes findings from the young research discipline of gelotology as a starting point and investigates how social interactions with robots can generate and implement cheerful moods. The positive effects of laughter and humor on stress, health and immune functions have been demonstrated in various studies. These results are currently being discussed in psychoneuroimmunological research [cf. 15, 16], which deals with the interactions between the nervous system and the immune system and their effects on well-being and disease development [cf. 17]. Since the 1960s, research has focused on the influence of positive emotions on health and well-being. The proposed project assumes that positive emotions can be generated through social interactions with robots. This strengthens the relationship between humans and robots and leads to a higher acceptance of robot systems.

From the perspective of the nursing sciences, there is an insufficient data situation regarding the effects of the use of nursing robotics stated and it is clearly discernible that the specific characteristics of professional nursing as a personal service have so far hardly been adequately addressed [3]. This indicates that the aim of care robotics in Europe to date has not been to address the specific needs, needs and particularities in the various fields of care (inpatient acute care, inpatient long-term care, outpatient care), but rather to test technological solution options for social problems in a way that is not specific to each field. An important task for nursing and nursing science is to emphasize the specific problems and peculiarities in nursing and to demand and support corresponding developments and evaluations and to introduce them into the discourse [cf. 3]. This is taken up by the SOHO project, in which it chooses a field of action-specific approach and makes inpatient care the starting point for the development of robotic solutions.

3.2   Research Questions

The SOHO project focuses on the following research questions on the level of "human", "organization" and "technology":

Human: Subjective satisfaction: To what extent are client and caregiver satisfied with the robotic solution? Does the robotic health assessment in the care process provide more time for the professionals to give their attention? Interaction: How does the use of the robotic system change interpersonal contact? Does the humanoid appearance stimulate the conversation with clients better than non-anthropomorphic systems? Does the robotic system change the feeling of loneliness? Are salutogenic effects of human-robot interaction identifiable? Qualification: What qualification and training requirements are associated with the use of the robotic system? What requirements does the robotic system place on the clients?

Organization: Adaptation/further development of Health Assessment procedures: Which existing methods can be used or further developed to record the subjective well-being of the client (PROMs) using a robotic system? Effect on care teams and care quality: What support and relief potential does the robotic system actually offer clients and caregivers? How do complex work processes change through its use in nursing? Implementation: Which supply settings will have to be set up in the future in order to organize the use of the robotic system in care? Financing: Which business models are suitable for using robotic solutions in practice?

Technology: Natural Language Analysis: How can a health assessment be supported by in-stream language analysis? Equipment: What technological features must the robotic system have? To what extent is a manipulator necessary? Technical implementation: Which procedures and algorithms are suitable for the implementation of assistant functions in maintenance? Operation and maintenance: What skills and effort are required to operate and maintain such systems?

By addressing these research questions on a concrete robotic system, the SOHO project contributes to the fact that future assistance robots will enable more quality of life and care for people with illnesses, disabilities and age-related restrictions.

3.3 Development Process

The humanoid robots of the Pepper type (Softbank Robotics), which are available for experiments, should be able to record the mood or state of mind of patients and determine their behavior (compliance with medication, drinking behavior, etc.) by means of various assessments. Nurses are to be supported by a Health Assessment Report generated on the basis of this data with a higher level of information in order to be able to draw conclusions about the condition of the patient and to have time for more intensive interactions with the patient by outsourcing the data collection.

The present findings reveal the importance of interdisciplinary cooperation between nursing practice, nursing science and technical sciences in the development and implementation of robot care systems. To address this need, the development process in the SOHO project is based on the User Experience Design (UX Design) approach and is supported by a highly interdisciplinary and transdisciplinary cooperation with patients, caregivers and experts from the health sector in order to achieve the above-mentioned project goals [18]. UX Design is designed to provide the user with the best possible experience when operating products or services. The UX Design is especially responsible for the design of the man-machine interface. The development process consists of the following phases (Fig. 1):

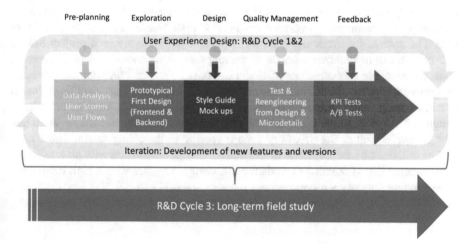

Fig. 1. SOHO development process

The designer usually begins with concepts that describe how the user can achieve his goals and use the planned functions efficiently (usability). Among other things, personas and user stories are defined, which are developed together with nursing experts and patients. It is shown for which users in which situations and in which way the planned robotic system should fulfil its purpose. Once the functionality has been clarified, UX Design continues to serve to create an offer that users like to use (Desirability or Joy of Use). In addition, the willingness to achieve testable results faster using UX rapid prototyping is gaining ground.

The SOHO project uses this approach in a process consisting of three cycles. In the first cycle we will develop a first demonstrator and test look and feel and basic functions together with patients and caregivers. In the second cycle the demonstrator will be improved to be ready for the third cycle where we will test the system in long term evaluation in the relevant environment. The added value that the SOHO project generates consists in particular in the development of urgently needed solutions in the field of care robotics and the holistic development perspective (user-centered approach, business model development and exploitation partner), which aims at the transfer potential of the solution to be developed. Moreover, there is still far too little empirical data in the field of care robotics to show to what extent these systems offer effective support. Through a formative and summative evaluation of the robotic system the project generates new insights in the field of nursing science and in the field of social robotics. The SOHO project uses transdisciplinary research to generate social innovations together with practice in creative collaboration and thus contributes to overcoming classical approaches of "research about practice" and to increasingly establishing modern participatory research.

4   Conclusion

In this work, we propose an approach for the development of a socio-assistive system for nursing care. For people who need care service, their quality of life and care will be improved by using such a system. For the nursing staff, their burden will get relieved as well, hopefully resulting in opportunities in an overall improvement of the quality of care. The state-of-the-art analysis shows in what aspects the research project SOHO will contributes to improve robotics in nursing care. This complex field of research holds considerable economic potential and can only be realized through intensive and integrative research activities in a pre-competitive environment. The scientific-technical risk lies among others in the question of the acceptance and feasibility of the robotic system with respect to the complex and essential functionalities needed by the target user groups as well as the necessary actions specific in the caring sector. These risks will be mitigated by the expertise of the SOHO project consortium with an integrative R&D process that is oriented towards the UX design, together with a high motivation in developing not only a robotic system but also practical and effective accompanying processes leading to sustainable contributions to the advancement of the care sector.

References

1. Bundesministerium für Gesundheit: Beschäftigte in der Pflege - Pflegekräfte nach SGB XI – Soziale Pflegeversicherung [online]. http://www.bmg.bund.de/themen/pflege/pflegekraefte/pflegefachkraeftemangel.html (2018). Accessed 15 Feb 2019

2. Wissenschaftliches Institut der AOK: Wenn der Beruf krank macht [online]. http://aokbv.de/imperia/md/aokbv/presse/pressemitteilungen/archiv/2015/wido_pm_krstd_2015-03-31.pdf (2015). Accessed 15 Feb 2018

3. Hülsken-Giesler, M., Daxberger, S.: Robotik in der Pflege aus pflegewissenschaftlicher Perspektive. Springer Gabler, Wiesbaden (2018)

4. Mitchell, P.H., Ferketich, S., Jennings, B.M.: Quality health outcomes model. Image J. Nurs. Sch. **30**, 43–46 (1998)

5. Drach, L.M., Terner, B.: Einsamkeit im Alter - Gesundheitsrisiko und therapeutische Herausforderung. Psychotherapie im Alter **36**, 441–457 (2012)

6. Bemelmans, R., Gerlderblom, G.J., Jonker, P., de Witte, L.: Socially assistive robots in elderly care: a systematic review into effects and effectiveness. J. Am. Med. Dir. Assoc. **13** (2), 114–120.e1 (2010)

7. Louie, W.-Y.G., McColl, D., Nejat, G.: Acceptance and attitudes toward a human-like socially assistive robot by older adults. Assistive Technol. **26**(3), 140–150 (2014)

8. Parobots, PARO Therapeutic Robot [online]. http://www.parorobots.com/ (2017). Accessed 15 Dec 2017

9. Robinson, H., MacDonald, B., Broadbent, E.: Physiological effects of a companion robot on blood pressure of older people in residential care facility: a pilot study. Australas. J. Ageing **34**(1), 27–32 (2015)

10. Büro für Technikfolgen-Abschätzung beim Deutschen Bundestag (TAB): Sachstandsbericht zum TAProjekt "Mensch-Maschine-Entgrenzungen: zwischen künstlicher Intelligenz und Human Enhancement" [online]. https://www.tab-beim-bundestag.de/de/pdf/publikationen/berichte/TAB-Arbeitsbericht-ab167.pdf (2016). Accessed 15 Feb 2019

11. Schwenk, A., Yuan, C.: Visual perception and analysis as first steps toward human-robot chess playing. In: Bebis, G. et al. (eds.) Advances in Visual Computing, LNCS. vol. 9475, pp. 283–292, Springer, Berlin/Heidelberg (2015)

12. Schaal S.: Max-Planck-Institut für intelligente Systeme, Tübingen, Roboter werden selbstständig [online]. https://www.mpg.de/9269151/jahresbericht-2014-schaal.pdf (2014). Accessed 15 Feb 2019

13. Bedaf, S., Gelderblom, G.J., de Witte, L.: Overview and categorization of robots supporting independent living of elderly people: what activities do they support and how far have they developed. Assistive Technol. **27**(2), 88–100 (2015)

14. Krings, B.-J., Böhle, K., Decker, M., Nierling, L., Schneider, C.: ITA-Monitoring „Serviceroboter in Pflegearrangements" [online] http://www.itas.fzk.de/deu/lit/epp/2012/krua12-pre01.pdf (2012). Accessed 15 Feb 2019

15. Martin, R.A.: Is laughter the best medicine? humor, laughter, and physical health. Curr. Dir. Psychol. Sci. **11**(6), 216–220 (2002)

16. Schor, J.: Emotions and health: laughter really is good medicine. Nat. Med. J. **2**(1), 1–4 (2010)

17. Bennett, M.P., Lengacher, C.A.: Humor and laughter may influence health. I. History and background. Evid. Based complement. Altern. Med. **3**(1), 61–63 (2006)

18. Hershey, P.: On the basis of the UX process for product design. https://www.uxbooth.com/articles/ux-a-process-or-a-task/. Accessed 21 Feb 2019

Understanding the Preference of the Elderly for Companion Robot Design

Suyeon Oh[1], Young Hoon Oh[2], and Da Young Ju[1(✉)]

[1] Technology and Design Research Center,
Yonsei Institute of Convergence Technology,
Yonsei University, Incheon, South Korea
{ochul21, dyju}@yonsei.ac.kr
[2] School of Integrated Technology,
Yonsei Institute of Convergence Technology,
Yonsei University, Incheon, South Korea
50hoon@yonsei.ac.kr

Abstract. Companion robots have been utilized as technical solutions to alleviate the problems of the elderly. Various types of companion robots have been suggested by human–robot interaction researchers. However, the appearance of some companion robots lacks analysis from the formative perspective. Therefore, we developed five robot design concepts based on literature reviews and the formative analysis of commercialized robots. In total, 19 participants of different age groups were interviewed to rate their preference for the developed design concepts. Cross-tabulation analysis and qualitative findings showed that the preference for the design concepts differed with the age group. The elderly preferred rounded and anthropomorphic robot designs. When evaluating the appearance of a robot, they preferred an intimate design enabling talking to the robot. However, the younger adults preferred neat and tidy designs with less detailed design elements. They considered the actual usage of the robot, emphasizing on maintenance and sanitation.

Keywords: Companion robot · Elderly · Appearance · Design concept

1 Introduction

The elderly feel less comfortable with new technologies such as robots than youngsters [1, 2]. To alleviate their concern, it is important for elder care robots to have an attractive appearance. It is known that visual aesthetics influence the perceptions of a user, including the first impression of a product, in several ways [3]. In the field of human–robot interaction, the appearance and morphology of a robot are also known to be important factors [4]. An agreeable appearance might increase the use of a robot and its interaction with the users. For instance, in a study, the elderly participants interacted with a cat robot more than the young ones because its appearance was small and non-intimidating, which met their preference [5].

Several studies have investigated how the elderly perceive the appearance of companion robots [6–9]. Most studies analyzed the preference of old adults based on

© Springer Nature Switzerland AG 2020
J. Chen (Ed.): AHFE 2019, AISC 962, pp. 92–103, 2020.
https://doi.org/10.1007/978-3-030-20467-9_9

categorization (e.g., anthropomorphic design, zoomorphic design, and machine-like design), and their research scope was limited to either commercial robots or unsophisticated design robots for the appearance categories. Not all the robots were designed for the elderly, so that they may not reflect their essential preferences. Therefore, we considered the preference of elderly individuals for robot appearance and developed our own robot concept design. This might assist in overcoming the shortcomings of the previous studies.

In this study, the aim of the research was to investigate the preference of the elderly for companion robots. We developed five robot design concepts based on: (i) the preference of the elderly from a literature review and (ii) the analysis of commercial robot design trends. Regarding the design concepts, we interviewed two age groups: young (aged from the 30s to 40s) and old (aged from the 50s to 70s). A comparative analysis of the two age groups was conducted to clearly draw insights about the elderly. The cross-tabulation analysis results and qualitative findings showed that the elderly preferred round shaped and anthropomorphic designs, whereas young adults preferred a neat and tidy robot design having less detailed elements. This paper contributes to deepen the understanding of the robot preferences of the elderly.

2 Literature Review

2.1 Companion Robot

A companion robot is commonly defined as a robot that can provide companionship or robot-assisted therapy to human beings [10]. Syrdal et al. believed that a companion robot should have two essential functions: (1) be useful and (2) behave socially [11]. A companion robot can perform useful tasks in a socially acceptable manner. This is also consistent with the definition in [2].

Studies have reported the various benefits of a companion robot. First, companion robots can be a better substitute of animal-assisted therapy [12–15]. This is because companion robots can alleviate concerns such as allergies and sanitation. Moreover, human–robot interaction researchers have reported that a companion robot could improve the psychological well-being of the elderly by the same extent as a real pet [16]. Other studies also found that therapy with a seal robot reduced the stress and increased the social communication of the elderly [17].

2.2 Appearance of the Companion Robot

Previous studies have showed that the design of a robot appearance plays an important role in human–robot interaction. Appearance is one of the important factors that affects robot acceptance [8]. The perception of the appearance of a robot may vary with age. The elderly spend more time and enjoy interacting with a small cat robot more than younger adults [18]. Even though the elderly have higher technological barriers compared to younger individuals [1], it seems that small-sized and non-intimidating designs lower the barrier [5]. The relationship between the age of a user and his/her perception of the appearance of a robot has been commonly reported in other studies.

Three to five years old children have felt frightened of android and humanoid robots, whereas infants tended to be attracted to them [19].

Other studies reported that the resemblance of a robot to a human being influenced the perceptions for them. Prakash et al. observed distinctive different preferences in human-likeness to a robot appearance based on the age group of the user [20]. Young adults had a relatively higher preference for machine-like robots, whereas the elderly had a higher preference for human-like robots. The authors suggested that these distinctive trends may be attributed to the experience with robots. These findings are somewhat consistent with [8]. Wu et al. found that the elderly showed positive attitude toward creative small robots with human traits such as eyes and mouth [8]. However, these previous studies might have some limitations. The robot design samples in them limited to commercial robots and research platforms. In addition, not all the robots were developed based on the characteristics of the elderly. Regardless of the features or usefulness of the robot, analysis of the appearance of an elderly care robot from a morphological perspective is required.

3   Design Concept of Companion Robot for the Elderly

In this section, we describe the formative characteristics of commercial companion robots. We categorized conventional robots and developed concept designs based on the analysis of the aforementioned characteristics. Design concepts were also designed on the basis of the findings from the literature regarding the appearance of companion robots.

3.1   Design Trend of Commercial Companion Robot

To analyze the formative aspects of general companion robots, we examined various types of robots. Moreover, we grouped them into three categories based on their shape attribute as: (1) simple, (2) iconic, and (3) character.

First, a simple type robot is made of a simple mass, and most of the conversational agents are in this group, such as the Amazon Echo. The formative characteristic of the simple type robots is that they are composed of one huge mass and have a relatively simple shape. Thus, there is a limitation in expressing emotions through gestures or movements owing to a lack of specific elements such as arms or legs. Moreover, less friendliness can be shown to be attributed to the inexistence of a notable characteristic as opposed to other types.

The second type is the iconic type. This type is made from a combination of simple shapes, e.g., Jibo [21], Kuri, and Zenbo, and it mostly consists of two parts: head and body. The iconic type, similar to the simple type, is simple in shape, but it has a distinct head and body. Unlike the following character type, however, this type of robot either has fixed limbs or no limbs, so that it is unable to interact with users through gestures or postures.

Last, the character type has visual traits similar to the individual nature of an animal or a human. This type of robot has a distinctive difference from the above two types in that it has movable elements such as limbs and head. Consequently, this type of robot can interact with a user through various gestures by moving the arms or legs. However, users may feel concerned when a robot highly resembles a human being or an animal.

3.2   Design Concept of the Companion Robot for the Elderly

As mentioned above, we investigated various design cases of commercial companion robots and categorized them further into five design concepts: (a) simple rabbit, (b) iconic rectangle, (c) iconic circle, (d) character teddy bear, and (e) simple rectangle. As it can be deduced from their names, (a) and (e) are the simple type, (b) and (c) belong to the iconic type, and (d) is classified as the character type. All the design concepts have an embedded display to print various facial expressions and play media content. These design concepts (Fig. 1) are rendered by various tools such as Rhinoceros 3D and Unigraphics NX6.

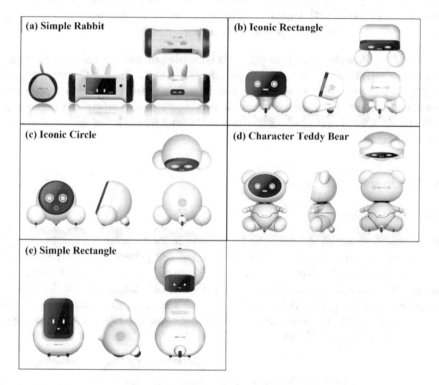

Fig. 1. Five design concepts used in the study, (a) simple rabbit, (b) iconic rectangle, (c) iconic circle, (d) character teddy bear, and (e) simple rectangle

(a) Simple Rabbit. (a) is composed of one single mass, which has no distinguishable between body and head. Unlike other concepts, (a) is long sideways so that there is less possibility of plunging and higher stability comparatively. We have added a characteristic design element, e.g., rabbit ears, to the simple type so that the users can feel a familiarity. The ears are designed to be detachable and made of silicone. We investigate the views of the subjects on these design elements.

(b) Iconic Rectangle. (b) consists of a head and body. (b) has a rectangular display, which can profit from minimizing the unnecessary bezel space. Thus, the screen area where the content is actually displayed can be maximized. To avoid the danger of injury, all the four corners are rounded.

(c) Iconic Circle. (c) as well as (b) consists of a head and body. (c) is composed of a repetitive combination of spherical elements. Unlike (b), (c) has a circular head design, similar to a human head or an animal head, which will increase the friendliness of the robot when conveying emotions and facial expressions.

(d) Character Teddy Bear. (d) is based on a teddy bear. In addition, (d) is designed based on the survey result that the elderly prefer teddy bear robots the most among animal shaped robots [22]. Unlike other design concepts, (d) has distinctive arms, legs, and ears, all of which can move similar to actual animals.

(e) Simple Rectangle. Similar to (a), (e) is composed of a single mass. (e) belongs to the simple type category like (a), but (e) adopts a portrait display. Consequently, the display is placed relatively high compared to in (a), which helps users with a more comfortable viewing experience when viewing down from a height. To improve the stability, the robot is designed with an inflated bottom part to lower the position of the center of mass.

4 Methods

4.1 Participants

In this study, we aimed to investigate the preference for the proposed design concepts. To analyze the relationship between the user age and preferred design concept of a robot, we involved participants aged from the 30s to the 70s. We divided them into two age groups: young (aged from the 30s to the 40s) and old (from the 50s to the 70s), and compared each two group. A total of 19 subjects was engaged for this study (female n = 11). Table 1 lists the demographic information of the participants. The mean age of the participants was 51.57 years.

Table 1. Demographic information of the participants

Age	Male(%)	Female(%)	Total(%)
30s	2 (10.5%)	2 (10.5%)	4 (21.2%)
40s	1 (5.3%)	3 (15.8%)	4 (21.2%)
50s	2 (10.5%)	1 (5.3%)	3 (15.8%)
60s	2 (10.5%)	4 (21.1%)	6 (31.6%)
70s	1 (5.3%)	1 (5.3%)	2 (10.5%)
Total	8 (42.1%)	11 (57.9%)	19 (100%)

4.2   Procedure

All the participants were briefed about the research purpose and concept of companion robots [10, 11]. They signed their informed consent and gave their demographic information in the questionnaire. All the five concept designs were printed on A4 sized sheets respectively (Fig. 2). The participants were given the sheets, and they responded in the interview by observing the five design concepts: (a) simple rabbit, (b) iconic rectangle, (c) iconic circle, (d) character teddy bear, (e) simple rectangle. They examined the design concepts for a sufficiently long time.

Fig. 2. Interview with the elderly

Semi-structured interviews were conducted to investigate their preferences for the five design concepts. They were asked to share their impressions on what they liked the most and the reasons for it. They also expressed the order of their preferred design concepts. All the surveys and interviews were conducted at their home or a cafe near their home (on the request of the participants). All the interviews were voice recorded with the consent of all the participants and were transcribed verbatim.

5   Result

5.1   Cross-Tabulation Analysis

In this section, we present the statistical relationship between the preference for a design concept and age group of the participant. By comparing the young and old groups, we aimed to extend our understanding of the needs of the elderly to gain more clarity. The dependent variables were the preference ranks of the robot design concepts (1st, 2nd, 3rd, 4th, and 5th). For instance, '1st' refers to the favorite design concept of a participant among the companion robots, '2nd' denotes his/her second most preferred design concept, and so on. The relationship was examined by cross-tabulation and Fisher's exact test. In all the tests, a p-value less than 0.05 was considered statistically significant.

5.2   Design Concept Preference by Age Group

The cross-tabulation analysis, as presented in Table 2, shows that there is a difference in the preference for the design concepts of companion robots based on the age group. According to Table 2, the young group prefers (e) the most, whereas the old group prefers (d) the most. None of the old group participants respond to (e) as their favorite concept design. Because more than 80% of the cells have expected values less than 5, we use Fisher's exact test. The test confirms that their favorite design concepts are significantly different ($p < 0.05$, Table 3). However, there is no significant difference for the 2nd, 3rd, 4th, and 5th choices based on the age group ($p > 0.05$).

Table 2. Cross-tabulation analysis: age group and 1st design concept

Age group		Design concept (1st)					
		(a)	(b)	(c)	(d)	(e)	Total
Young	Count	0	1	2	2	3	8
	% within young group	0%	12.5%	25.0%	25.0%	37.5%	100.0%
Old	Count	0	1	1	9	0	11
	% within old group	0%	9.1%	9.1%	81.8%	0.0%	100.0%
Total	Count	0	2	3	11	3	19
	% with in Total	0%	10.5%	15.8%	57.9%	15.8%	100.0%

Table 3. Fisher's exact test: age group and 1st design concept

	Value	Sig.
Pearson Chi-Square	7.501	0.040*
Likelihood ratio	8.841	0.53
Fisher's exact test	7.245	0.022*
No. of valid cases	19	

5.3   Interview Result

(a) Simple Rabbit. (a) was the least preferred design concept by the young group. Most of them responded that they examined whether a design was sufficiently practical for use. In this regard, they disliked its rabbit ear because they considered it as ineffective and merely an embellishment (P8, P14, P16, P17, and P18). They believed that it would be hard to place in a narrow area due to the shape (P17). Moreover, there was a response regarding the small display, making the user experience uncomfortable (P14).

"*It seems that it has a small display. Even if it has a larger display, it would be better to not have the ears... they might hinder the movement of the robot... it may have been suitable if it was a table clock but not as a robot...*" - *P14 (young group, 30s)*

"*I think it will be difficult to place it anywhere owing to its shape. Other designs appear as if they can be placed in a narrow area, but this is not for (a). It looks extremely fragile. If it was to move around, it would most likely break. It does not feel like a companion robot.*" - *P17 (young group, 30s)*

The elderly shared both positive and negative views on the appearance of (a). When they focused on the facial features of (a) (P2, P10, and P11), they responded that the design seemed to be disproportionate, but they considered the rabbit ears were appealing and familiar. They expressed that the overall shape resembled a machine, with which they would not talk to (P1, P3, and P15). Moreover, the square shape of (a) was reminiscent of an outdated TV design (P1, P5, and P9).

"This (a) looks extremely rigid like a machine. However, the ears are appealing. It has a type of attractiveness from the rabbit ears. Looking closer, I can see it has eyes and a nose..." - P2 (old group, 70s)

"The forehead of (a) is too wide. I think the ears are appealing, but they are inappropriate with respect to the overall shape of the robot. The rabbit like ears are not matching with its sleepy eyes. Rabbits should have round eyes." - P11 (old group, 50s)

"It is like electronics. It feels like a machine. Old TVs looked just like this. Somehow the overall shape gives a strong industrial impression. I feel like having a conversation with (d), but not with (a). Why? Because it is just a machine." – P15 (old group, 50s)

(b) Iconic Rectangle. The young group showed a negative attitude toward (b) because they thought it is consisted of unnecessary design elements (P13, P14, P16, and P17).

"What I mean by numerous elements is that there are simply too many things happening. I thought that the elements are making the robot unstable. Moreover, it would be difficult to clean all the dust collecting everywhere possible. At first glance, I thought (b) and (c) looked similar, but after a second viewing, (b) is not as attractive as (c) exterior-wise." - P17 (young group, 30s)

They also commonly mentioned that the appearance of (b) seemed unstable (P8, P16) and that its ratio is inappropriate (P17 and P19). They also answered that it would be inconvenient for a user to carry or clean the robot (P14 and P17).

"(b) looks awkward. It might be different with a real product, but for example the unbalanced, huge head appears like an apple and a circle in the same place." - P19 (young group, 40s)

The old group thought (b) had an 'unfriendly design'. They preferred round shapes to rectangular ones and responded that the rectangular design elements from (b) were reminiscent of old-school designs in the past (P15). Moreover, they mostly focused on human-like aspects when evaluating the design of (b) (P3, P8, P10, and P11).

"The shape of (b) seems as if it belongs to my generation. In the past, all the electronics were rectangular. Consider car designs; they were first rectangular, and then became curvy. People tend to reject things from the past because they do not want to recall old days." - P15 (old group, 50s)

"I do not believe that I would be able to have a conversation with (b). It simply seems to be like a machine and does not appear friendly at all. Regarding the face and shape, I cannot see its facial features clearly. Its eyes look fatigued..." – P3 (old group, 60s)

(c) Iconic Circle. The young group showed a positive attitude toward (c) because it appeared to be stable and agreeable. None of them particularly disliked the design.

The old group preferred its rounded designs (P2, P4, P9, P10, P11, and P15), and they focused on the facial features of (c) such as eyes and mouth (P1, P4, P10, and P11).

"This I think, looks stable. This has three wheels and move like a robot. I like it." - *P6 (young group, 40s).*

"(c) has human eyes and round lips, making it look like a human face. It looks nice..." - *P10 (old group, 60s)*

(d) Character Teddy Bear. Most of the young group negatively responded toward (d). They thought that its limbs were non-functional and useless (P13, P14, P16, and P17).

"To me, (d) appears somewhat unstable... It looks like (d) is made up of several parts. I doubt that it will be easy to maintain. As you can see, the linking parts seem extremely fragile, as if will very easily break if the robot falls on the ground. In addition, if dirt collects between the parts, it will not look appealing visually." - *P16 (young group, 30s)*

Contrastingly, the old group responded that the "Character shaped robots are friendly and appealing" (all the old group) and preferred (d) the most (all the old group, except P8 and P12).

"This seems the most appealing and the most stable because it is in a sitting position. (d) is attractive and feels friendly. The bear shape It looks familiar. The others are slightly odd, but this looks right in my eyes. It resembles the toys for children, which makes it friendly. The round shape also makes it seem friendlier." - *P2 (old group, 70s)*

(e) Simple Rectangle. The young group preferred (e), particularly its "simplicity". Considering the usage of a companion robot, the young group mentioned that this robot would be actually easier to maintain or clean compared to other design concepts by being one single mass.

"I thought (e) was better because the other designs have numerous elements... I thought that its design as one complete entity looked easy to clean, and so I chose type (e). Type (d) would become easily dirty." - *P17 (young group, 30s)*

"Well, it serves the purpose of a robot. Some design elements are added from conversational agents. I like that (e) has some unique features." - *P19 (young group, 40s)*

Contrastingly, the old group showed a negative attitude toward (e). Mostly, they responded that "It is a less intimate design and is reminiscent of a mechanical product."

"Compared to (e), the other designs have their own shape or form, such as a teddy bear, making them friendlier. However, (e) simply makes an observer question what it is. So, (e) scores negatively in terms of friendliness, and (e) is also not exactly stable. Its eyes look like a power outlet, which seems a bad design." - *P8 (old group, 60s)*

6 Discussion

Our findings showed that the old group preferred the rounded and human-like appearance of the companion robots [20]. They commonly examined the existence of facial feature and a natural appearance. These qualitative findings may imply that the elderly prefer character type companion robots [22], which have human-like facial features [8, 20].

In addition, the old group tended to consider the rectangle-shaped concepts as electronic products. They felt it would be awkward to talk with such robots, whereas it would be natural to talk to the character type robot, (d), because it was appealing and familiar. This may be owing to the fact that the elderly prefer familiar objects [23].

The young group preferred the simple type design, (e), which had few external design elements. However, they disliked the character type, (d), because it was composed of many irrelevant parts. In line with this trend, they also showed a negative attitude toward simple type (a), which had an additional animal-like characteristic, i.e., nonfunctional rabbit ears. Moreover, they were concerned about the issues of actually using a robot, such as its durability and maintenance. This was another reason why the young group disliked (d), whose parts would make it difficult for the users to manage it.

In conclusion, we found significant differences in the preferred design concepts with age. The elderly with low technology acceptance or concerns about new technology [1, 2] preferred designs familiar to them. Contrastingly, the young people with relatively more experience of using new technologies [24], tended to emphasize on the practicality of management and usability of the product rather than its friendly appearance.

This study has several limitations. First, this study lacks specific results on how much the participants preferred each design concept. Our cross-tabulation analysis shows the relationship between the preference for a design concept and age, but shortcomings may still exist in regard to the comparison of the preferences for the different designs in each design concept category. Second, the number of respondents was limited to 19, which might limit the statistical analysis. In the future work, it will be necessary to engage more participants and conduct experiments to improve the reliability of the results. Last, we used robot pictures [8, 20] to examine the impressions made regarding the appearance, but this process might not have been sufficient to determine how users perceive the design of a robot. If we were given actual robots, then the interviews would have been able to investigate more practical impressions.

7   Conclusion

The goal of this research was to investigate the preferences of senior citizens for companion robots. Based on the analysis of the formative characteristics of conventional commercial robots and characteristics of the elderly, we devised five design concepts of companion robots. Using design concept images, we interviewed a wide range of age groups from the 30s to the 70s about the appearance of the robots. On analyzing the preferences based on the age, the old group tended to prefer familiar, anthropomorphic designs. They preferred the character type and responded that it had a friendly appearance. Contrastingly, the young group showed a preference for a simple and clean design; this was because they regarded that features without functionalities are cause inconvenience to the users. The findings of this research showed that the user age was relevant for designing companion robots and could contribute to improve companion robot designs such that they were visually acceptable to the elderly.

Acknowledgments. This research was supported by Research Program to Solve Social Issues of the National Research Foundation of Korea (NRF) funded by the Ministry of Science and ICT (NRF-2017M3C8A8091770). This research was supported by the MIST (Ministry of Science and ICT), Korea, under the "ICT Consilience Creative Program" (IITP-2018-2017-0-01015) supervised by the IITP (Institute for Information & Communications Technology Promotion). We would like to thank Jimin Ryu for designing the companion robot images.

References

1. Czaja, S.J., Sharit, J.: Age differences in attitudes toward computers. J Geron. Ser. B. **53B**, 329–340 (1998)
2. Schraft, R.D., Schaeffer, C., May, T.: Care-O-bot: the concept of a system for assisting elderly or disabled persons in home environments. In: IECON '98. Proceedings of the 24th Annual Conference of the IEEE Industrial Electronics Society, pp. 2476–2481, IEEE (1998)
3. Brunel, F.F., Bloch, P.H., Arnold, T.J.: Individual differences in the centrality of visual product aesthetics: concept and measurement. J. Consum. Res. **29**, 551–565 (2003)
4. Del Pobil, A.P., Sundar, S.S.: Interaction science perspective on HRI: designing robot morphology. In: Proceedings of the 5th ACM/IEEE International Conference on Human-robot Interaction, p. 5. IEEE Press, Piscataway, NJ, USA (2010)
5. Broadbent, E., Stafford, R., MacDonald, B.: Acceptance of healthcare robots for the older population: review and future directions. Int. J. Soc. Robot. **1**, 319 (2009)
6. Syrdal, D.S., Dautenhahn, K., Woods, S.N., Walters, M.L., Koay, K.L.: Looking good? appearance preferences and robot personality inferences at zero acquaintance. In: AAAI Spring Symposium: Multidisciplinary Collaboration for Socially Assistive Robotics (2007)
7. Li, D., Rau, P.L.P., Li, Y.: A cross-cultural study: effect of robot appearance and task. Int. J. Soc. Robot. **2**, 175–186 (2010)
8. Wu, Y.-H., Fassert, C., Rigaud, A.-S.: Designing robots for the elderly: appearance issue and beyond. Arch. Gerontol. Geriatr. **54**, 121–126 (2012)
9. Walters, M.L., Koay, K.L., Syrdal, D.S., Dautenhahn, K., Te Boekhorst, R.: Preferences and perceptions of robot appearance and embodiment in human-robot interaction Trials. In: Procs New Front. Human-Robot Interact (2009)
10. Heerink, M., Kröse, B., Evers, V., Wielinga, B.: Assessing acceptance of assistive social agent technology by older adults: the Almere model. Int. J. Soc. Robot. **2**, 361–375 (2010)
11. Syrdal, D.S., Dautenhahn, K., Woods, S., Walters, M.L., Koay, K.L.: 'Doing the right thing wrong'-personality and tolerance to uncomfortable robot approaches. In: Robot and Human Interactive Communication, 2006. ROMAN 2006. The 15th IEEE International Symposium on. pp. 183–188. IEEE (2006)
12. Preuß, D., Legal, F.: Living with the animals: animal or robotic companions for the elderly in smart homes? J. Med. Ethics. **43**, 407 LP-410 (2017)
13. Banks, M.R., Willoughby, L.M., Banks, W.A.: Animal-assisted therapy and loneliness in nursing homes: use of robotic versus living dogs. J. Am. Med. Dir. Assoc. **9**, 173–177 (2008)
14. Friedmann, E., Galik, E., Thomas, S.A., Hall, P.S., Chung, S.Y., McCune, S.: Evaluation of a Pet-assisted living intervention for improving functional status in assisted living residents with mild to moderate cognitive impairment: a pilot study. Am. J. Alzheimer's Dis. Other Dementias®. **30**, 276–289 (2014)

15. Shibata, T.: Therapeutic seal robot as biofeedback medical device: qualitative and quantitative evaluations of robot therapy in dementia care. Proc. IEEE **100**, 2527–2538 (2012)
16. Shibata, T., Wada, K., Tanie, K.: Statistical analysis and comparison of questionnaire results of subjective evaluations of seal robot in Japan and UK. In: 2003 IEEE International Conference on Robotics and Automation, pp. 3152–3157. IEEE (2003)
17. Yu, R., Hui, E., Lee, J., Poon, D., Ng, A., Sit, K., Ip, K., Yeung, F., Wong, M., Shibata, T.: Use of a therapeutic, socially assistive pet robot (PARO) in improving mood and stimulating social interaction and communication for people with dementia: study protocol for a randomized controlled trial. JMIR Res. Protoc. **4**, e45 (2015)
18. Libin, A.V., Libin, E.V.: Person-robot Interactions from the robopsychologists' point of view: The robotic psychology and robotherapy approach. Proc. IEEE **92**, 1789–1803 (2004)
19. Minato, T., Shimada, M., Ishiguro, H., Itakura, S.: Development of an android robot for studying human-robot interaction BT - innovations in applied artificial intelligence. Presented at the (2004)
20. Prakash, A., Rogers, W.A.: Why some humanoid faces are perceived more positively than others: effects of human-likeness and task. Int. J. Soc. Robot. **7**, 309–331 (2015)
21. Breazeal, C.: Social robots: from research to commercialization. In: Proceedings of the 2017 ACM/IEEE International Conference on Human-Robot Interaction, p. 1. ACM (2017)
22. Oh, Y.H., Kim, J., Ju, D.Y.: Analysis of design elements to enhance acceptance of companion robot in older adults. In: RO-MAN 2018 - The 27th IEEE International Symposium on Robot and Human Interactive Communication, Workshop on Social Cues in Robot Interaction, Trust and Acceptance. IEEE (2018)
23. Lazar, A., Thompson, H.J., Piper, A.M., Demiris, G.: Rethinking the design of robotic pets for older adults. In: Proceedings of the 2016 ACM Conference on Designing Interactive Systems, pp. 1034–1046. ACM (2016)
24. Lee, S., Choi, J.: Enhancing user experience with conversational agent for movie recommendation: effects of self-disclosure and reciprocity. Int. J. Hum Comput Stud. **103**, 95–105 (2017)

A Study on the Design of Companion Robots Preferred by the Elderly

Soo Yeon Kim[1], Young Hoon Oh[2], and Da Young Ju[3(✉)]

[1] Creative Technology Management, Underwood International College,
Yonsei University, Seoul, South Korea
sykim213@yonsei.ac.kr
[2] School of Integrated Technology,
Yonsei Institute of Convergence Technology,
Yonsei University, Incheon, South Korea
50hoon@yonsei.ac.kr
[3] Technology and Design Research Center,
Yonsei Institute of Convergence Technology, Yonsei University,
Incheon, South Korea
dyju@yonsei.ac.kr

Abstract. As the elderly population grows rapidly, companion robots have attracted attention as a technological solution for problems faced by the elderly. Although the design elements of companion robots have influenced their usability, there remains a lack of relevant research. Therefore, this paper aims to extend our understanding of the design elements of companion robots preferred by the elderly. We conducted experiments on what types, weight, and materials are preferred for the elderly aged 50–64. Furthermore, we analyzed whether there were statistically significant differences in design preferences according to gender, living arrangement, and age. Consequently, the preference of the female elderly for robots over 2 kg was observed to be significantly lower. Additionally, elderly people living alone preferred synthetic fur over elderly people living with others. In conclusion, it was found that the specific weight and material of the companion robot affects the preference of elderly people.

Keywords: Companion robot · Human-Robot interaction · Elderly · Design · User experience

1 Introduction

The rapid growth in the elderly population is a global problem. Internationally, the elderly population is growing faster than the youth population [1]. As the elderly population increases, it is expected that more elderly care services will become necessary in society [2]. As the health of elderly people deteriorates, they will be less likely to be self-sufficient and require more nursing homes and social welfare facilities. However, the current social welfare system cannot accommodate the rapidly growing elderly population; thus, the elderly community will require subsidiary equipment for support [3].

© Springer Nature Switzerland AG 2020
J. Chen (Ed.): AHFE 2019, AISC 962, pp. 104–115, 2020.
https://doi.org/10.1007/978-3-030-20467-9_10

The elderly population faces various social problems. One of the most serious social problems is depression and loneliness [4]. It is found that loneliness may cause an depression in an elderly person [5], and the mortality due to depression is very high [6]. Many studies state that elderly people living alone feel the most loneliness [7, 8]. In previous studies, elderly people that live alone report much higher levels of depression than elderly people living with their families or spouses [9, 10]. This suggests that living alone has a significant influence on psychological distress in elderly people. Furthermore, because depression in the elderly is highly associated with suicide or suicidal thoughts [11], loneliness among the elderly population should be considered as a critical problem.

Companion robots are emerging in the form of technology that can solve the problem of depression in elderly people [12]. A companion provides all the advantages of pets without the drawbacks. Companion robots provide positive physiological effects (improvement of biological signals), psychological effects, and social impact, all of which are provided by pets [13–15]. In contrast to companion robots, pets may cause a variety of problems, e.g., causing injuries to the elderly and causing parasitic infections. Raising a pet at a nursing home or care facility requires significant attention. On the other hand, companion robots do not need to be fed or bathed [13, 14].

Moreover, studies have found that companion robots can provide psychological stability to the elderly and foster sociality [16]. In a 5 year case study of robot therapy, it was observed that the companion robot PARO positively influenced 14 elderly patients who suffered from mild dementia in nursing homes in Japan [17]. PARO relieved the elderly's stress and brought a soothing atmosphere to the nursing homes. In addition, another experiment shows that companion robots can improve brain function by reducing stress hormones in elderly people [18, 19]. In the experiment, two companion robots, namely AIBO and PARO, were brought into a care facility for elderly residents. The elderly group interacted with the robots for over 9 h each day. The elderly residents actively interacted with PARO and experienced positive changes in their stress levels. In the case of the AIBO robot, a U.S research group conducted an experiment to determine whether robotic dogs were effective in relieving loneliness in elderly people [20]. The elderly group spent 30 min weekly with AIBO for 8 weeks. The results indicated that AIBO was effective in decreasing loneliness in elderly people that live in long-term care facilities.

With advances in the field of robotics, companion robots are programmed to behave, speak, and interact with humans [21]. Many HRI researchers are immersed in implementing various functions in companion robots. Although prior research focused on the feasibility and usefulness of companion robots, there is still a lack of research focused on the design elements of companion robots. The design elements of companion robots (e.g. shape, material, and weight of the robot) are important to increase user acceptability. In order to determine the design elements suitable for companion robots, the type, weight, and material preferred by elderly group should be examined. In this study, an experiment was conducted with a group of elderly people wherein their preferences were determined based on age, gender, and living arrangement.

This paper is organized as follows. Prior studies on companion robots for use by the elderly will be reviewed in Sect. 2. An experimental method and the derived result will be explained in the following Sects. 3 and 4. Implications and limitations of this study will be discussed in Sects. 5 and 6.

2   Literature Review

2.1   Companion Robots for Elderly Emotional Health

As the number of elderly people suffering from loneliness and depression grows, HRI researchers are eager to develop companion robots that can help solve these problems [22]. It was found that elderly people feel relieved when they socialize with a companion robot [23]. In Germany, an experiment was conducted with 16 elderly people. The elderly living with family or spouses were aged from 50 to 75, and the elderly living in a nursing home were aged from 75 to 98. Both groups spent time with the NAO robot; NAO sang a song for the elderly, read the newspaper, and talked with the elderly group. The result was very positive in that both elderly groups felt relieved when they socialized with the NAO robot. Thus, companion robots are in demand by lonely elderly people, even those with advanced age. Like the prior studies suggest, companion robots seem to be highly necessary for the elderly because these robots can provide psychological stability to the elderly and foster sociality [18, 19].

2.2   Acceptance of a Companion Robot by the Elderly

In the near future, companion robots will be commonly used by the elderly. However, many people cast doubt on the acceptance of companion robots by the elderly. It is true that the elderly group are rarely willing to accept new technology that might require time and effort [24]. In Amsterdam, there was an interesting experiment where iCat was used to confirm the researchers' hypothesis regarding the acceptance of a companion robot. They stated that the elderly could develop social abilities while interacting with iCat, and this would contribute to their sense of social presence. Thus, this experience would provide the elderly with greater enjoyment than any other activities, thereby increasing the acceptance level [25]. There were 30 participants aged from 65 to 94 who partially lived alone. The results show that the elderly's acceptance of companion robots was high when they felt enjoyment after interacting with iCat. The experiment highlighted the importance of the social presence experience of the elderly group.

3   Method

3.1   Participants

This study involved 15 elderly participants. Unlike a previous study [26], we conducted an experiment with elderly people under the age of 65. Nine of the 15 respondents were recruited from a re-education institution for senior citizens over the age of 50, and the remaining six were selected through the snowball sampling method. The mean age of the participants was 57.93 years old (SD = 4.667). All respondents were residents of Seoul. Table 1 shows the demographic information of the recruited participants. After the experiment, respondents received a $30 gift card as compensation.

Table 1. Participant demographic information.

Demographics	Data (%)
Age: M (SD)	57.93 (4.667)
Gender: n (%)	
Male	3 (20.0%)
Female	12 (80.0%)
Living arrangement: n (%)	
Living alone	4 (26.7%)
Living with others	11 (73.3%)

3.2 Questionnaire

The questionnaire included the following items:
  Section A: Demographic information (gender, age, and living arrangement);
  Section B: Design elements (type, weight, and material of companion robots).

3.3 Procedure

Before the experiment, we briefly explained the purpose of the survey to the participants. All participants responded to Section A after providing informed consent. Before they responded to Section B, we explained the definitions and concepts of companion robots to all participants [27, 28]. Regarding Section B, experiments involving design aspects were conducted in the following order: type, weight, and material of the companion robot. The content of the experiment matched that in a previous study [26].

Type. We used four stuffed toys and one toy robot for respondents in the Type experiment (bear, seal, dog, humanoid, and newborn baby, as shown in Fig. 1). The five types of toys did not operate or sound like an actual robot, and they were used exclusively to investigate the preference for a certain type of companion robot. Animal robots were chosen based on the representative morphology of robots frequently used in previous studies (e.g., bear [29], seal [30], dog [20], and humanoid [31]). In addition to types of animals, research on humanoids robot has been conducted for several years; thus, we included humanoid robots in this study. We selected two robots in human shape, one as a humanoid and another as a newborn baby, to compare the preferences towards a humanoid robot and an animal robot.

Fig. 1. Stuffed animal toys and a humanoid robot used in the Type test (from left, bear, seal, dog, humanoid, newborn baby).

We presented five types of robots to the respondents. We asked them to consider those robots as companion robots. Every respondent answered a questionnaire based on the five-point Likert scale, where each robot was evaluated by touching and holding the robot for a sufficient amount of time.

Weight. The appropriate weight of a companion robot was not thoroughly investigated in previous studies. Thus, we attempted to investigate preferences towards various weights. We took a stuffed dog and attached a wrist sandbag for weight experiment. As dogs are the most common pet in South Korea, we used a stuffed dog as a representative sample. The weight of a stuffed dog without a wrist sandbag (default) is 150 g; we attached an additional wrist sandbags to increase its weight to 650 g, 1.15 kg, 1.65 kg, 2.15 kg, and 2.65 kg (Fig. 2). We examined the elderly's preference to different weights. The respondents received guidance regarding whether this dog was a companion robot and evaluated the robot on a 5-point Likert scale for each weight.

Fig. 2. Stuffed dogs are used in the Weight test (from left, 150 g, 650 g, 1.15 kg, 1.65 kg, 2.15 kg, 2.65 kg).

Material. To investigate the preferences for materials of companion robots, we used four representative materials for the experiment. We selected materials based on the following criteria. First, we selected materials that allow respondents to intuitively distinguish the type of material. Therefore, we considered materials that users can easily find in their daily lives. Because materials vary in their thickness and type of fiber, we chose a material with specific texture. Second, we selected materials based on previous research. Many previous studies described the importance of soft materials and texture [30, 32], thus we selected the material considering its softness. Based on these criteria, we selected silicone, microfiber, synthetic fiber, and plastic. All materials were clearly distinguishable (Fig. 3). Those materials could be easily seen in everyday life: Silicone - Apple iPhone Silicone Case, Plastic - Plastic package of iPhone Silicone Case, Microfiber - blanket, and Synthetic fiber - sleeping socks.

To examine the preferences for these materials, we created a 'Material test panel' (Fig. 3). We cut black hardboard paper to create four equally sized windows and placed four materials on the back of the hardboard. We controlled the contact area so that respondents touched these four materials on the windows. We requested that those

Fig. 3. Material panel is used in the Material test (from left, silicone, plastic, synthetic fibers, and microfibers).

materials be considered as the materials used to build the companion robot. All respondents provided a rating for each material on the five-point Likert scale after touching the four materials for a sufficient amount of time.

3.4 Statistical Analysis

All statistical analyses were conducted with SPSS (version 25.0). We classified all respondents into two groups based on age (50s (n = 7) and 60s (n = 8)), depending on gender, and depending on living arrangement (living alone (LA from below) and living with others (LWO from below)). Thus, the independent variables were gender, living arrangement, and age group. Due to the small number of participants, we used a non-parametric method (Mann-Whitney test) to compare the two independent groups. We investigated whether there were statistically significant differences in the design elements of the companion robots preferred by the elderly based on gender, type of living arrangement, and age group.

4 Results

4.1 Type

Gender. Unlike the results of a previous study [26], respondents in this experiment hold the greatest dislike for the Newborn Baby among the 5 types of companion robots (Table 2). For all types of companion robot, the Mann-Whitney tests do not indicate any significant difference in preference by gender (Table 3).

Living Arrangement. The LA group preferred Humanoids over the LWO group, but these preferences were not significantly different (Table 3). Seal was the least preferred type among animal robots (Table 2). The Mann-Whitney test results do not show any significant difference between the LA and LWO groups (Table 3).

Age Group. Regardless of age, the preference for the Newborn Baby companion robot was lowest (Table 2). Although the most preferred robot was different for the two age

Table 2. Type preference by gender, living arrangement, and age group.

Type	Gender		Living arrangement		Age group	
	Male	Female	Living alone	Living with others	50s	60s
Bear	4.00	4.08	4.50	3.91	4.00	4.13
Seal	3.33	3.33	3.25	3.36	3.29	3.38
Dog	4.33	3.75	4.25	3.73	3.43	4.25
Humanoid	4.00	3.75	4.75	3.45	3.57	4.00
Newborn baby	3.67	2.92	3.25	3.00	3.14	3.00

Table 3. Type preference – Mann-Whitney test results.

Type	Gender		Living arrangement		Age group	
	U	p-value	U	p-value	U	p-value
Bear	16.5	0.812	11.5	0.131	25.0	0.702
Seal	17.5	0.941	22.0	1.0	26.0	0.812
Dog	11.0	0.275	15.0	0.324	14.0	0.080
Humanoid	16.5	0.821	8.5	0.066	22.0	0.468
Newborn baby	11.5	0.333	18.5	0.637	27.0	0.905

groups (Table 2), there was no significant difference between 50s and 60s for all types of companion robot (Table 3). The 50s group favored Bear the most, while 60s favored Dog the most.

4.2 Weight

Gender. The most preferable weight of the companion robot was different according to gender. Males responded that 1.15 kg was the best weight among the six weights (Table 4). When the weight of the robot increased from 150 g to 1.15 kg, the preference for the robot also increased. On the contrary, females answered that 650 g was

Table 4. Weight preference by gender, living arrangement, and age group.

Weight	Gender		Living arrangement		Age group	
	Male	Female	Living alone	Living with others	50s	60s
150 g	2.67	3.25	3.25	3.09	2.57	3.63
650 g	3.33	3.50	2.50	3.82	3.86	3.13
1.15 kg	4.00	2.83	2.75	3.18	3.43	2.75
1.65 kg	3.33	1.92	1.50	2.45	2.29	2.13
2.15 kg	3.00	1.17	1.00	1.73	1.86	1.25
2.65 kg	2.00	1.00	1.00	1.27	1.43	1.00

Table 5. Weight preference – Mann-Whitney test results.

Weight	Gender		Living arrangement		Age group	
	U	p-value	U	p-value	U	p-value
150 g	15.0	0.650	20.5	0.838	18.5	0.250
650 g	15.5	0.710	11.0	0.139	18.5	0.257
1.15 kg	8.5	0.159	17.5	0.546	19.5	0.312
1.65 kg	6.0	0.064	11.0	0.125	27.0	0.902
2.15 kg	7.0	0.041*	14.0	0.179	25.0	0.655
2.65 kg	6.0	0.003*	18.0	0.377	20.0	0.118

Note. *$p < 0.05$.

the best among the six weights. As the weight of the robot increased to 1.15 kg, their preference for companion robots began to decline. Regardless of gender, the preference for 2.65 kg (heaviest weight) was the lowest.

Moreover, the Mann-Whitney test revealed a significant difference between 2.15 kg ($U = 7.0$, $p = 0.041$) and 2.65 kg ($U = 6.0$, $p = 0.003$, Table 5). This implied that men could accommodate robots with heavier weight than females.

Living Arrangement. The LA group responded that they most preferred 150 g, while the LWO group responded that they most preferred 650 g. Regardless of living arrangement, the preference for weight levels above 1.65 kg continued to decline (Table 4). The Mann-Whitney test results do not show any significant difference between the LA and LWO groups for all weight levels (Table 5).

Age Group. The most preferable weight for the 50s and 60s groups was different. The 50s and 60s groups replied that 650 g and 150 g were the most appropriate weights, respectively (Table 4). Additionally, there was a difference in priority levels of weight levels in the 50s and 60s. The preference for weights over 150 g decreased for the 60s group, but the 50s group preferred 1.15 kg over 150 g. This indicates that the 50s group thought that a companion robot should not be too light. Moreover, it revealed that the 60s group was more likely to reject the feeling of a heavy companion robot than the 50s group. However, the Mann-Whitney test results did not show a significant difference between the 50s and 60s groups (Table 5).

4.3   Material

Gender. Both males and females preferred microfiber materials, but the materials they disliked the most were different. Males most disliked synthetic fiber, and females most disliked plastic (Table 6). However, there was no significant difference in material preference for all materials (Table 7).

Living Arrangement. Regardless of living arrangement, all respondents preferred microfiber material and had the lowest preference for plastic (Table 6). We found a significant difference regarding the preference of synthetic fiber by living arrangement.

Table 6. Material preference by gender, living arrangement, and age group.

Material	Gender		Living arrangement		Age group	
	Male	Female	Living alone	Living with others	50s	60s
Silicone	4.33	3.50	3.50	3.73	3.71	3.63
Microfiber	4.33	4.25	4.75	4.09	4.71	3.88
Synthetic fiber	2.33	3.33	4.00	2.82	2.86	3.38
Plastic	2.67	1.75	1.75	2.00	2.00	1.88

Table 7. Material preference – Mann-Whitney test results.

Material	Gender		Living arrangement		Age group	
	U	p-value	U	p-value	U	p-value
Silicone	14.0	0.542	19.5	0.731	27.5	0.951
Microfiber	15.5	0.689	14.5	0.278	16.5	0.140
Synthetic fiber	6.5	0.080	6.5	0.033*	19.5	0.300
Plastic	7.5	0.107	18.5	0.627	26.0	0.806

Note. *p < 0.05.

The LWO group significantly disliked synthetic fiber over the LA group (U = 6.5, p = 0.033). The LWO group seemed to be more concerned about hygiene than the LA group (Table 7).

Age Group. Preferable material tendencies were similar regardless of age group. Both groups most preferred microfiber and they most disliked plastics. The next most preferable material was silicone, followed by synthetic fiber and plastic (Table 6). In addition, there were no notable differences in material preferences depending on age group (Table 7).

5 Discussion

Previous studies have shown various benefits of companion robots for the elderly. Companion robots reduced loneliness and depression in the elderly [4, 22]. Companion robots also provided comfort, relief, and psychological stability to elderly people who needed sincere caring [18]. These results represent the necessity of companion robots as a mediator in aging populations. To develop more adequate companion robots, we should consider which design elements of companion robots preferred by the elderly. Our results show that elderly women had significantly lower preference for a companion robot weighing more than 2 kg compared to men. In addition, the LA group significantly preferred a synthetic fur compared to the LWO group.

Regardless of gender, the preference for 2.65 kg (heaviest weight) was the lowest (Table 4). The Mann-Whitney test revealed a significant difference for 2.15 kg (p < 0.05) and 2.65 kg (p < 0.05, Table 5). This indicates that men could afford heavier robots than females. Females have significantly decreased preference when the weight of the robot exceeded 2 kg. All respondents most preferred microfiber, whereas the preference for plastic was lowest (Table 6). We found a significant difference regarding the preference of synthetic fiber by living arrangement. The LWO group significantly disliked microfiber over the LA group ($p < 0.05$). The LWO group seemed to be more concerned about hygienic aspects than the LA group.

There are some limitations in this study. First, the experimental results are limited to five kinds of robots. If size is more prominent and diverse, and given robots are more human-like or animated, the results could be different. Second, statistical analysis is limited because the number of respondents was limited to 15 people. Third, there were some restrictions on the interpretation of the survey results because there was no qualitative analysis.

6 Conclusion

This research was conducted to identify which design elements of companion robots are preferred by the elderly. We examined the preferences of elderly people of up to 65 years of age towards various types, weights, and materials of companion robots. We statistically analyzed whether there are significant differences in preferences for various designs by classifying 15 elderly people based on age, sex, and living arrangement. A female elderly person's preference decreased significantly when the weight of the robot exceeded 2 kg. The preference for 2.65 kg (heaviest weight) was the lowest, regardless of gender. All respondents preferred microfiber, while the preference for plastic was lowest. The LWO group significantly disliked microfiber compared to the LA group. These findings may aid HRI designers and developers to improve companion robots for the elderly.

Acknowledgement. This research was supported by Research Program To Solve Social Issues of the National Research Foundation of Korea (NRF) funded by the Ministry of Science and ICT (NRF-2017M3C8A8091770). This research was supported by the MIST (Ministry of Science and ICT), Korea, under the "ICT Consilience Creative Program" (IITP-2018-2017-0-01015) supervised by the IITP (Institute for Information & communications Technology Promotion). We are particularly grateful for the assistance given by Jaewoong Kim and Su Wan Park.

References

1. United Nations: World Population Ageing 2017 Highlights, New York (2017)
2. Jacobzone, S.: Coping with aging: international challenges. Health Aff. **19**, 213–225 (2000). https://doi.org/10.1377/hlthaff.19.3.213
3. Barba, B., Tesh, A., Courts, N.: Promoting thriving in nursing homes: the Eden Alternative. J. Gerontol. Nurs. (2002). https://doi.org/10.1136/bmjopen-2014-004916
4. Djernes, J.K.: Prevalence and predictors of depression in populations of elderly: a review. Acta Psychiatr. Scand. **113**, 372–387 (2006). https://doi.org/10.1111/j.1600-0447.2006.00770.x

5. Cacioppo, J.T., Hughes, M.E., Waite, L.J., Hawkley, L.C., Thisted, R.A.: Loneliness as a specific risk factor for depressive symptoms: cross-sectional and longitudinal analyses. Psychol. Aging **21**, 140–151 (2006). https://doi.org/10.1037/0882-7974.21.1.140

6. Stek, M.L., Vinkers, D.J., Gussekloo, J., Beekman, A.T.F., van der Mast, R.C., Westendorp, R.G.J.: Is depression in old age fatal only when people feel lonely? Am. J. Psychiatry. **162**, 178–180 (2005). https://doi.org/10.1176/appi.ajp.162.1.178

7. Mui, A.C.: Depression among elderly Chinese immigrants: an exploratory study. Soc. Work **41**, 633–645 (1996). https://doi.org/10.1093/sw/41.6.633

8. Lim, L.L., Kua, E.-H.: Living alone, loneliness, and psychological well-being of older persons in Singapore. Curr. Gerontol. Geriatr. Res. **2011**, 673181 (2011). https://doi.org/10.1155/2011/673181

9. Chou, K.-L., Chi, I.: Comparison between elderly Chinese living alone and those living with others. J. Gerontol. Soc. Work **33**, 51–66 (2000). https://doi.org/10.1300/J083v33n04_05

10. Russell, D., Taylor, J.: Living alone and depressive symptoms: the influence of gender, physical disability, and social support among Hispanic and non-Hispanic older adults. J. Gerontol. Ser. B Psychol. Sci. Soc. Sci. **64B**, 95–104 (2009). https://doi.org/10.1093/geronb/gbn002

11. Cavanagh, J.T.O., Carson, A.J., Sharpe, M., Lawrie, S.M.: Psychological autopsy studies of suicide: a systematic review. Psychol. Med. **33**, 395–405 (2003)

12. Kachouie, R., Sedighadeli, S., Khosla, R., Chu, M.T.: Socially assistive robots in elderly care: a mixed-method systematic literature review. Int. J. Hum. Comput. Interact. (2014). https://doi.org/10.1080/10447318.2013.873278

13. Preuß, D., Legal, F.: Living with the animals: animal or robotic companions for the elderly in smart homes? J. Med. Ethics (2017). https://doi.org/10.1136/medethics-2016-103603

14. Friedmann, E., Galik, E., Thomas, S.A., Hall, P.S., Chung, S.Y., McCune, S.: Evaluation of a pet-assisted living intervention for improving functional status in assisted living residents with mild to moderate cognitive impairment: a pilot study. Am. J. Alzheimers. Dis. Other Demen. (2015). https://doi.org/10.1177/1533317514545477

15. Shibata, T.: Therapeutic seal robot as biofeedback medical device: qualitative and quantitative evaluations of robot therapy in dementia care. In: Proceedings of the IEEE (2012)

16. Shibata, T., Wada, K.: Robot therapy: a new approach for mental healthcare of the elderly - a mini-review (2011)

17. Wada, K., Shibata, T., Kawaguchi, Y.: Long-term robot therapy in a health service facility for the aged - a case study for 5 years. In: 2009 IEEE International Conference on Rehabilitation Robotics, ICORR 2009 (2009)

18. Wada, K., Shibata, T.: Living with seal robots - its sociopsychological and physiological influences on the elderly at a care house. In: IEEE Transactions on Robotics (2007)

19. Wada, K., Shibatal, T., Musha, T., Kimura, S.: Effects of robot therapy for demented patients evaluated by EEG. In: 2005 IEEE/RSJ International Conference on Intelligent Robots and Systems, IROS (2005)

20. Banks, M.R., Willoughby, L.M., Banks, W.A.: Animal-assisted therapy and loneliness in nursing homes: use of robotic versus living dogs. J. Am. Med. Dir. Assoc. **9**, 173–177 (2008). https://doi.org/10.1016/J.JAMDA.2007.11.007

21. Lucidi, P.B., Nardi, D.: Companion robots: the hallucinatory danger of human-robot interactions five ethical concerns for companion robots (2018)

22. Miehle, J., Bagci, I., Minker, W., Ultes, S.: A social companion and conversation partner for elderly

23. Bemelmans, R., Gelderblom, G.J., Jonker, P., de Witte, L.: Socially assistive robots in elderly care: a systematic review into effects and effectiveness. J. Am. Med. Dir. Assoc. **13**, 114–120.e1 (2012). https://doi.org/10.1016/j.jamda.2010.10.002
24. Forlizzi, J., DiSalvo, C., Gemperle, F.: Assistive robotics and an ecology of elders living independently in their homes. Human-Computer Interact. (2004). https://doi.org/10.1207/s15327051hci1901&2_3
25. Heerink, M., Kröse, B., Evers, V., Wielinga, B.: The influence of social presence on acceptance of a companion robot by older people
26. Oh, Y.H., Kim, J., Ju, D.Y.: Analysis of design elements to enhance acceptance of companion robot in older adults. In: RO-MAN 2018 - The 27th IEEE International Symposium on Robot and Human Interactive Communication, Workshop on Social Cues in Robot Interaction, Trust and Acceptance (2018)
27. Heerink, M., Kröse, B., Evers, V., Wielinga, B.: Assessing acceptance of assistive social agent technology by older adults: the almere model. Int. J. Soc. Robot. **2**, 361–375 (2010). https://doi.org/10.1007/s12369-010-0068-5
28. Sharkey, A.: Robots and human dignity: a consideration of the effects of robot care on the dignity of older people. Ethics Inf. Technol. **16**, 63–75 (2014). https://doi.org/10.1007/s10676-014-9338-5
29. Moyle, W., Jones, C., Sung, B., Bramble, M., O'Dwyer, S., Blumenstein, M., Estivill-Castro, V.: What effect does an animal robot called CuDDler have on the engagement and emotional response of older people with dementia? a pilot feasibility study. Int. J. Soc. Robot. **8**, 145–156 (2016). https://doi.org/10.1007/s12369-015-0326-7
30. Kidd, C.D., Taggart, W., Turkle, S.: A sociable robot to encourage social interaction among the elderly
31. Gouaillier, D., Hugel, V., Blazevic, P., Kilner, C., Monceaux, J., Lafourcade, P., Marnier, B., Serre, J., Maisonnier, B.: Mechatronic design of NAO humanoid. In: 2009 IEEE International Conference on Robotics and Automation (2009)
32. Heerink, M., Albo-Canals, J., Valenti-Soler, M., Martinez-Martin, P., Zondag, J., Smits, C., Anisuzzaman, S.: Exploring requirements and alternative pet robots for robot assisted therapy with older adults with dementia. Presented at the October 27 (2013)

Human-Robot Communications and Teaming

Assessment of Manned-Unmanned Team Performance: Comprehensive After-Action Review Technology Development

Ralph W. Brewer II[1]([⊠]), Anthony J. Walker[2], E. Ray Pursel[3],
Eduardo J. Cerame[4], Anthony L. Baker[1], and Kristin E. Schaefer[1]

[1] US Army Research Laboratory, Aberdeen Proving Ground,
Aberdeen, MD, USA
{ralph.w.brewer.civ,
kristin.e.schaefer-lay.civ}@mail.mil,
tlbaker27@gmail.com
[2] DCS Corporation, Alexandria, VA, USA
ajwalker@dcscorp.com
[3] Naval Surface Warfare Center, Dahlgren, VA, USA
eugene.pursel@navy.mil
[4] US Army TARDEC, Warren, MI, USA
eduardo.j.cerame.civ@mail.mil

Abstract. Training in the US Army starts with the individual. Soldiers work on acquiring skills, knowledge, and attributes in order to perform tasks to support operational requirements. Feedback is provided through After-Action Reviews (AARs) to support training and improve future operations. A main difficulty for developing effective training for manned-unmanned teams (MUM-T) is that AARs with human-agent teams are yet to be developed. While AAR processes for human teams are well trained in the Army, the current methods for delivering an AAR do not account for unmanned systems that are integrated in collective tasks. The US Army's Robotic Wingman program provides a use case for discussing potential technology solutions that can support critical human factors for MUM-T during a gunnery collective task. Understanding the capabilities and limitations of an unmanned platform will help develop effective training plans and performance measures for the unmanned asset which is now part of the team.

Keywords: Manned-unmanned teaming · Robotic · Wingman ·
After-Action Review · Human-agent teaming · Autonomous systems

1 Introduction

Training is the cornerstone to producing a cohesive and proficient fighting force and After-Action Reviews (AARs) are a key part of that process. The U.S. Army's AAR can trace its roots back to the post World War II era. However, the AAR methods in use today came about in the mid 1970's [1]. For evaluating modern combat, AARs provide feedback on mission and task performances in training and combat to correct

This is a U.S. government work and not under copyright protection in the U.S.;
foreign copyright protection may apply 2020
J. Chen (Ed.): AHFE 2019, AISC 962, pp. 119–130, 2020.
https://doi.org/10.1007/978-3-030-20467-9_11

deficiencies, sustain strengths, and focus on performance of specific training objectives once a mission is complete [2]. During the AAR each member of the team provides input to evaluate the mission and provide insight into what to sustain or how to correct failures for future missions. Traditionally, the AAR facilitator seeks maximum participation from the crew using open-ended questions to keep the discussion dynamic, candid, and professional[1].

While it takes some training to become an effective facilitator [3], most leaders are aware of the purpose and the process of the AAR for human teams. AARs are conducted during or directly following the event. The AAR must focus on the training objectives of the event and the Soldier, leader, and unit performance. The AAR determines what the strengths and weaknesses are of the unit and the Soldiers. The results of the AAR will determine training or retraining for subsequent events.

A main difficulty for developing effective training for manned-unmanned teams (MUM-T) is that AARs with human-agent or human-autonomy teams are yet to be developed. The difficulty lies in the fact that at least one member is an unmanned system and cannot verbally communicate in the AAR discussion. For these types of teams, effective teaming and appropriate use of the technology is dependent on the human's understanding of the system, its behaviors, and the reasoning behind those behaviors. The AAR must transform as the members of the unit change. Prior to all this induction of technology into the Army, exercise data was collected and manipulated manually [4]. In a group training atmosphere there is an issue of generating proper feedback for the team members, but the addition of technology into the AAR process allows added insight into distributed decision-making and knowledge of the team [5]. The use of computer aided AAR tools can provide ground truth data to couple with the perceived truth of the human members of the team [6].

2 Background: AARs and Human-Agent Teaming

When looking into technology-based solutions for improving AARs for MUM-T, it is important to retain the key principles of an AAR. These technologies must allow team members to engage in debriefs in order to integrate lessons learned from past performance [7]; provide opportunities to practice resolving conflicts in a productive manner [8, 9] in order to improve coordination during future performance; allow team members to develop accurate mental models of problems, tasks, spaces, and shared knowledge [10, 11]; and provide a constructive process of reviewing performance to improve a team's ability to monitor their performance, recognize errors, and self-correct [12, 13]. Without some type of technology into the AAR process, it is almost impossible to integrate the unmanned asset as part of the team.

AAR technologies can support three major areas of human factors to enhance teaming: communication amongst the human crew and with an unmanned asset, shared situation awareness across the entire team, and team trust. To effectively meet the needs

---

[1] The Training Circular (TC) 25-20, A Leaders Guide to After-Action Reviews is the guide that the Army developed to help leaders plan, prepare, and conduct an effective AAR [2].

of modern MUM-T, these AAR technologies provide a method for observing and evaluating mission-specific, environmental, and social context. This is critical because an understanding of these contexts can improve team comprehension and understanding, thus improving performance and collaborative decision-making [14, 15]. In a similar vein, an AAR can help teams diagnose and understand their performance by providing context for outcomes and situations that occurred during the scenario.

2.1 Team Communication

Crew communication is integral to most team processes, and therefore to team performance [9]. Communication positively affects team performance by improving other critical team processes such as decision making, problem detection, and coordination. For example, it allows team members to resolve disagreements, synchronize information from multiple sources, and align goals [12]. Failures in communication are commonly linked to team performance failures in critical situations [16]. However, teams that communicate effectively are better able to perform under duress and deal with crisis situations [17].

Intelligent agents introduce complexities into the dynamics of human team communication. Human teammates use communication to coordinate behaviors, and can often anticipate information needs and deliver useful information to the correct team members when needed with minimal requests for coordination [18]. While these subtle coordination abilities are the target of active development among human-agent team (HAT) researchers [19], processes like naturalistic, human-like communication protocols are still being developed [20]. As a result, key challenges in collaborative HAT involve the capability of the agent to clearly communicate its intentions to human team members and the design of user displays to allow agents to communicate intent [21]. Given these challenges, AARs should provide human-agent teams with a mechanism for reviewing coordination and communication behaviors during simulated or live tasks, identifying errors in communication, and improving future coordination during HAT.

2.2 Shared Situation Awareness and Transparency

Shared situation awareness (SA) and effective transparency are critical aspects of HAT. Both factors relate to each team member's understanding of the task, themselves, their teammates, the mission goals, and the environment. When discussing AAR technologies to support MUM-T, transparency is defined as the degree to which the logic and processes used by an unmanned system are 'known' to human teammates [22]. This is critical because a transparent system supports both individual and team understanding related to system states, as well as insight into its reasons for making decisions and taking actions. This directly affects the development of both team and shared SA[2]. Having an appropriate amount of shared SA about the mission,

---

[2] Team SA encompasses the joint decisions and actions whereby each individual team member may have their own SA to carry out individual goals. Shared SA is the overlap of information that directly affects team coordination [11, 23].

environment, and social context, the human team member can better predict the other teammates' needs and act on those needs before a direct request for action [21], but a lack of information sharing and teamwork, and poor information system reliability can lead to incorrect SA for decision-making leading to a critical failure for any of the team members [24].

AAR technologies should directly support improved transparency and support both team and shared SA. Having a visual perspective of individual crew stations as well as a more global view of the collective task can improve perception of elements in the environment, comprehension of the situation, and better projection of future team operations, all elements which are critical to developing appropriate SA. More specifically, AAR technologies should provide a means to discuss both crew and unmanned teammate decision-making.

2.3   Team Trust

Trust has been identified as critical aspect of teamwork and has a consistent positive relationship with team performance for both human teams [25], and effective human-agent teams, especially during high risk operations that occur in uncertain and dynamic environments [26]. Trust between the crew serves to improve cohesion, satisfaction, and perceived effectiveness among human team members [27]; while trust of an unmanned system can impact team operations as well as directed use of that agent [28].

To advance team effectiveness related to trust, AAR technologies should support coordination for determining decision authority, better processes for managing trust-based decisions, and developing a shared mental decision model [29]. Ultimately, these types of technologies can help disambiguate differences between user perceptions and actual vehicle behaviors, thus help effectively calibrate trust in the team. AARs should also provide value to the crew (or human) dimension of trust. AARs allow teams to cooperatively diagnose their performance and engage in essential processes such as team self-correction, goal-setting, and conflict resolution [30].

3   Background: AARs for Gunnery Operations

Unmanned assets can integrate far more information from multiple sources faster than their human counterparts. They are potentially faster, cheaper, more persistent, and are more precise. It is important to effectively team humans and unmanned assets to take advantage of that increased lethality. AARs that include feedback from all members of this MUM-T will increase the efficiency of the team and improve performance.

3.1   Gunnery Performance

The US Army provides a set of standards for training and evaluating live fire gunnery operations for manned crews of direct fire ground systems with Training Circular (TC) 3-20.31 titled Training and Qualification, Crew [31]. The manual goes through the entire process of training, range operations, the engagement process, range requirements and scenario development, and the qualification standards for the crew tables [31]. It outlines the AAR process to provide a forum for Soldiers to discuss the events and discover what happened, why it happened, and how to sustain or improve on this in the future.

Crew table qualification takes place on a live fire range using a scenario that meets the standard set forth in the TC. The associated targets are hidden down range and raised up on lifting mechanisms based on the engagement scenario. Three specific types of targets are shown in Fig. 1 are examples of range targets for evaluating a crew using a vehicle mounted machine gun.

Fig. 1. Stationary truck (left), infantry (center), and moving truck (right) targets

A scenario consists of 10 engagements comprised of both offensive and defensive postures against stationary and moving targets during daylight and night operations. The targets are exposed for 50 s after target lock in which the crew must engage the targets as quickly as possible with their weapon system. Each engagement is worth a maximum of 100 points. The standard for qualification is to score at least 700 of 1000 points, whereby 70 points or more for all targets in 7 of 10 engagements, and qualify at least one night engagement.

Coordination within the crew starts with a fire command to reduce confusion and increase proficiency. The crew can then efficiently place accurate fires on the threats. The vehicle commander (VC) will begin by issuing an initial fire command and then follow up with a subsequent fire command if the same target should be reengaged or a supplemental fire command if moving to service another target. Responses required by the crew are: identified, on the way, cannot identify, and cannot engage. These commands establish a common conduct of fire within the crew and provide information to the crew of what actions are being taken.

3.2 Evaluating Gunnery Operations

An experienced and trained team of vehicle crew evaluators (VCEs) work in concert to assess the proficiency of each crew on their ability to negotiate the course during each of the ten engagements. During the engagements the VCEs record both video and audio of the run. The VCEs enter the timing of the engagement of targets into the common crew score sheet which is DA Form 8265 [31] to determine a base score of the engagement. The audio of the crew provides a picture of how the crew was working together during the run. The VCE will listen to the fire command, responses to that command, log any penalties and notate good and bad points from the engagement.

Once the crew has finished their run, they will meet the primary evaluator for an in-depth AAR. With the addition of an intelligent unmanned system to the crew, it is important to understand what that vehicle is seeing, thinking, and doing.

3.3   Typical Gunnery AAR

A typical gunnery evaluation AAR follows the general format outlined in TC 25-20 [32] whereby all participants are present and rules of engagement and training objectives are identified. The VCE asks open-ended questions of the crew, and examines performance through self-guided evaluation of each engagement in chronological order. For an AAR to be effective there has to be an accurate perception of the events [6]. If the crew has a shared mental model of the task, roles and interaction then they will be better able to anticipate the needs of one another [33]. Training and practice by the crew will allow this mental model to form and solidify. The crew has the perceived truth with the VCE providing the ground truth [23]. Current AAR technologies are limited to video and audio recordings of crew engagements. Having an autonomous unmanned asset as part of the crew adds a layer of complexity to discussion. Adjustments must be made to the AAR and how information is collected to support the process. The following use case provides insight into some options for addressing this area of need.

4   Wingman Use Case: Developing Technologies for MUM-T AAR

The US Army Wingman program consists of a team of manned and unmanned ground combat vehicles. Its goal is to increase the autonomous capabilities of the unmanned robotic combat vehicle (RCV; Fig. 2, left side) while coupling it with the manned command vehicle (MCV; Fig. 2, right side) to provide a lethal HAT. The crew of the MCV consists of a driver and VC, the robotic vehicle gunner (RVG) and the robotic vehicle operator (RVO) to support direct interaction with the RCV, and the long-range advanced scout surveillance system (LRAS3) operator to support target detection.

Fig. 2. Wingman vehicles

The Wingman RCV is a developed prototype system making this use case optimal to describe and demonstrate technologies developed to support effective AAR capabilities for future MUM-T. AAR capabilities include the capability to replay a mission with pertinent time stamps for critical events, understanding the location of each asset, threat locations, sensor feeds, external cues, verbal team communication, interactions with a user display, and the overall decision-making process of the RCV. Understanding the capabilities and limitations of an unmanned platform will help develop effective training plans and performance measures for the unmanned asset which is now part of the crew.

4.1 Technology Components that Support AARs

Technology provides a way to create AAR aids. The technology generated aids for the AAR should not be a replacement for the AAR leader [34]. It provides a tool set for the evaluator to help paint a picture for the crew which now includes the unmanned system. A key reminder for MUM-T AARs is that the unmanned asset is unable to speak for itself, so the data must articulate the story for the vehicle. This is critical for effective teaming to occur as appropriate use of the technology is dependent on the human's understanding of the system, its behaviors, and the reasoning behind those behaviors [2]. There are a few different ways that Wingman demonstrates how technology can be integrated into the AAR process: technology that supports the VCE, technology that supports the individual crew station, and technology that provides a global team perspective.

4.2 Technology Supporting the VCE

Within Wingman, there is a real-time video stream of the RVG's camera sensor and the crew audio provided to the tower for the VCE real-time observation to gather information, compile it, and use it to guide the AAR in an understandable format that is easy to follow for all crew stations. While these video and audio files are recorded and can be used to facilitate the AAR, they still do not provide a complete picture to the crew.

4.3 Individual Crew Station Technology Aids

Technology should support replay of the individual crew stations where possible. For Wingman, the crew interacts directly with the unmanned asset through the Warfighter Machine Interface (WMI) user display. The VC, RVG and RVO each have a version of the WMI which is set up to allow each to interact in their specific area of expertise (Fig. 3). The capability to replay the events from a mission for all crew members to observe may advance both an individual crew member's understanding of the events, but also be used to identify successes, failures, and potential system errors as a crew.

For this functionality to be possible, the WMI logger captures the values every widget and button press on the WMI every 100 ms. Each log file records multiple categories of information, the categories are parsable using a python script to separate out the data into individual files. But one major challenge involves parsing that

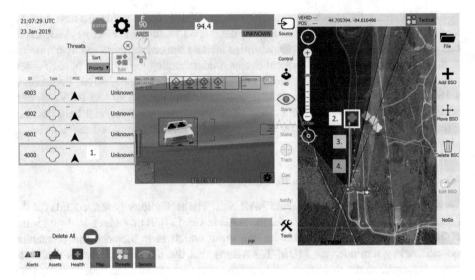

Fig. 3. Warfighter Machine Interface 1. Four potential targets in yellow (left side) with identified target in green box. 2. Identified target shown on map. 3. Arrow identifying the target is being tracked. 4. Waypoint path RVO entered for the RCV.

information and providing it in a useful format to facilitate growth of the individual and the crew. In addition, this playback type of feature is limited in that it only looks at a replay of those displays, not the entire social, environmental, or mission context. To reach that level of shared SA amongst the crew and the unmanned asset, a more global technology is needed.

4.4   Global Technology Aids

In the Wingman use case, AAR effectively allows the operators of the vehicles to review their performance from a more global, mission-based, top-down perspective. It allows operators to more clearly understand the outcomes of their actions and decisions by reviewing key aspects of the scenario. Critical to this process is the flow of information within the MCV, the RCV, and between the two to make decisions for the autonomous mobility and lethality. To reach this level of shared SA to improve communication, transparency and calibrate trust, a top-down replay of the world is needed (Fig. 4). This type of technology is possible from the WMI log file. A python script parses the recorded data into a timeline of events which is playable within Google Earth. Currently the timeline is able to show the location, position, and heading of each vehicle, the field of view of the LRAS3 sensor and the remote weapon system (RWS), and the robotic vehicle mobility state during the course of the run. It also shows the location of each of the LRAS3 operator's identified targets through a symbol on the map, as well as the actual target's geolocations.

Fig. 4. Global technology aids. 1. Battle space object (yellow) vs actual target. 2. LRAS3 operator Field of View (FOV) 3. Weapon FOV 4. RVO mobility path 5. Current vehicle locations and battle positions

Future work for this Wingman AAR technology will include data from other data collected from the software. First, data recorded from the Autonomous Remote Engagement System (ARES)[3] is crucial to understanding how the system supported autonomous engagement of targets during the exercise. By integrating these features onto this map, it will provide more information to the crew about the underlying decision-making process of the weapon system autonomy, but also by reviewing information logged via ARES, the crew can learn from their interactions with the system by identifying gaps in knowledge or training, diagnosing performance errors, and recognizing successful interactions.

Second, data collected from the Robotic Technology Kernel (RTK) autonomous mobility software, can be used to provide more accurate and specific data related to the vehicle position data (latitude, longitude, and elevation) coupled with the yaw, pitch, and roll of the weapon system. This data combined with the status messages of the gimbal (elevation and azimuth of the sight, range of target, tracking information, line of fire, and any error codes of the RWS), weapon (inhibit and error codes, burst rate and length), control handle (zoom, slew rate, focus, and trigger press), sight package, operator control (WMI), and ammunition provides a more complete view of the features that impact success or failure of each target engagement. Currently ROS has a feature to capture these messages as a bagfile which is playable in operating system for review. The goal of this ongoing effort is to provide rich interaction data to AAR participants, allowing them to review their interactions with the unmanned system and understand the antecedents to and consequences of their decisions and actions.

---

[3] ARES handles the decision making for the weapon system. Identification of a target in the field of view adds a red box to highlight the target, currently depicted on the RVG WMI.

5   Conclusion

Although MUM-T is a complex field of active development, the use of AARs can inform and improve future work in that domain. In this article, we have outlined the latest technologies that support AARs in the Wingman system, a cutting-edge use case for MUM-T. Information parsed from recorded user interactions with the RCV and with vehicle autonomy software can provide crews with a rich understanding of how they interacted with each other and with the unmanned system, providing a framework for addressing MUM-T specific issues and improving subsequent teamwork and SA. The increasing adoption of robotics and autonomous systems into Army contexts mirrors the ongoing efforts to augment AARs with new technologies, and the use case in the Wingman program provides an effective context for developing and deploying these new AAR methods. Development of a technology that can visualize data for the human to be able to parse and understand in the context of the training event is the ultimate goal. Turning the streaming data coming from the weapon system into useful information and collating that with the knowledge of the crew during an AAR will make the team safer, effective, and lethal.

Acknowledgment. The views and conclusions contained in this document are those of the authors and should not be interpreted as representing the official policies, either expressed or implied, of the Army Research Laboratory or the U.S. Government. The U.S. Government is authorized to reproduce and distribute reprints for Government purposes notwithstanding any copyright notation herein.

References

1. Morrison, J.E., Meliza, L.L.: Foundations of the After Action Review Process. Institute for Defense Analyses, Alexandria (1999)
2. Command, U.A.T.a.D.: A leader's guide to after-action reviews. In: (US), D.o.t.A. (ed.) Training Circular No.: TC 25-20, Washington, DC (1993 Sep 30)
3. DeGrosky, M.T.: Improving After Action Review (AAR) Practice. The Association (2005)
4. Allen, G., Smith, R.: After action review in military training simulations. In: Proceedings of Winter Simulation Conference, pp. 845–849. IEEE (1994)
5. Gratch, J., Mao, W.: Automating after action review: attributing blame or credit in team training. University of Southern California Marina Del Rey Information Sciences Institute (2003)
6. Johnson, C., Gonzalez, A.J.: Automated after action review: state-of-the-art review and trends. J. Defense Model. Simul. 5, 108–121 (2008)
7. Smith-Jentsch, K.A., Johnston, J.H., Payne, S.C.: Measuring team-related expertise in complex environments. Mak. Decisions Stress: Implications Individ. Team Training 1, 61–87 (1998)
8. De Dreu, C.K., Weingart, L.R.: Task versus relationship conflict, team performance, and team member satisfaction: a meta-analysis. J. Appl. Psychol. 88, 741 (2003)
9. Marks, M.A., Mathieu, J.E., Zaccaro, S.J.: A temporally based framework and taxonomy of team processes. Acad. Manag. Rev. 26, 356–376 (2001)

10. Blickensderfer, E., Cannon-Bowers, J.A., Salas, E.: Theoretical bases for team self-correction: fostering shared mental models. Adv. Interdisc. Stud. Work Teams **4**, 249–279 (1997)
11. Endsley, M.R.: Toward a theory of situation awareness in dynamic systems. Hum. Factors **37**, 32–64 (1995)
12. Salas, E., Sims, D.E., Burke, C.S.: Is there a "big five" in teamwork? Small Group Res. **36**, 555–599 (2005)
13. Marks, M.A., Panzer, F.J.: The influence of team monitoring on team processes and performance. Hum. Performance **17**, 25–41 (2004)
14. Schaefer, K.E., Aksaray, D., Wright, J.L., Roy, N.: Challenges with addressing context with AI and human-agent teaming. In: Lawless, W.F., Mittu, R., Sofge, D. (eds.) Computational Context: The Value, Theory and Application of Context with AI. CRC Press, Boca Raton (2019)
15. Schaefer, K.E., Oh, J., Aksaray, D., Barber, D.: Integrating context into artificial intelligence: research from the robotics collaborative technology alliance. AI Magazine (in press)
16. Sasou, K., Reason, J.: Team errors: definition and taxonomy. Reliab. Eng. Syst. Saf. **65**, 1–9 (1999)
17. Mckinney Jr., E.H., Barker, J.R., Davis, K.J., Smith, D.: How swift starting action teams get off the ground: what United flight 232 and airline flight crews can tell us about team communication. Manage. Commun. Quart. **19**, 198–237 (2005)
18. Cooke, N.J., Gorman, J.C., Myers, C.W., Duran, J.L.: Interactive team cognition. Cogn. Sci. **37**, 255–285 (2013)
19. McNeese, N.J., Demir, M., Cooke, N.J., Myers, C.: Teaming with a synthetic teammate: insights into human-autonomy teaming. Hum. Factors **60**, 262–273 (2018)
20. Bisk, Y., Yuret, D., Marcu, D.: Natural language communication with robots. In: Proceedings of the 2016 Conference of the North American Chapter of the Association for Computational Linguistics: Human Language Technologies, pp. 751–761 (2016)
21. Schaefer, K.E., Straub, E.R., Chen, J.Y., Putney, J., Evans III, A.W.: Communicating intent to develop shared situation awareness and engender trust in human-agent teams. Cogn. Syst. Res. **46**, 26–39 (2017)
22. Hoff, K.A., Bashir, M.: Trust in automation: Integrating empirical evidence on factors that influence trust. Hum. Factors **57**, 407–434 (2015)
23. Meliza, L.L., Goldberg, S.L., Lampton, D.R.: After action review in simulation-based training. Army Research IInst for the Behavioral and Social Sciences (2007)
24. Kaber, D.B., Endsley, M.R.: Team situation awareness for process control safety and performance. Process Saf. Prog. **17**, 43–48 (1998)
25. de Jong, B.A., Dirks, K.T., Gillespie, N.: Trust and team performance: a meta-analysis of main effects, contingencies, and qualifiers. In: Academy of Management Proceedings, p. 14561. Academy of Management Briarcliff Manor, NY 10510 (2015)
26. Schaefer, K.E., Chen, J.Y., Szalma, J.L., Hancock, P.A.: A meta-analysis of factors influencing the development of trust in automation: implications for understanding autonomy in future systems. Hum. Factors **58**, 377–400 (2016)
27. Costa, A.C.: Work team trust and effectiveness. Pers. Rev. **32**, 605–622 (2003)
28. Lee, J.D., See, K.A.: Trust in automation: designing for appropriate reliance. Hum. Factors **46**, 50–80 (2004)
29. Gremillion, G.M., Donavanik, D., Neubauer, C.E., Brody, J.D., Schaefer, K.E.: Estimating human state from simulated assisted driving with stochastic filtering techniques. In: International Conference on Applied Human Factors and Ergonomics, pp. 113–125. Springer (2018)

30. Salas, E., Rosen, M.A., Burke, C.S., Goodwin, G.F.: The wisdom of collectives in organizations: an update of the teamwork competencies. In: Team Effectiveness in Complex Organizations: Cross-Disciplinary Perspectives and Approaches, pp. 39–79. Routledge, New York (2009)
31. Command, U.A.T.a.D.: Training and qualification, crew. In: (US), D.o.t.A. (ed.) Training Circular No.: TC 3-20.31, Washington, DC (2015 Mar 17)
32. A leader's guide to after-action reviews. U.S. Army Training Circular 25-20 (1993)
33. Smith-Jentsch, K.A., Cannon-Bowers, J.A., Tannenbaum, S.I., Salas, E.: Guided team self-correction: impacts on team mental models, processes, and effectiveness. Small Group Res. **39**, 303–327 (2008)
34. Dyer, J.L., Wampler, R.L., Blankenbeckler, P.N.: After action reviews with the ground Soldier system. Army Research IInst for the Behavioral and Social Sciences (2005)

An Uncertainty Principle for Interdependence: Laying the Theoretical Groundwork for Human-Machine Teams

W. F. Lawless[(✉)]

Paine College, 1235 15th Street, Augusta, GA, Georgia
w.lawless@icloud.com

Abstract. The deliberateness of rational decision-making is an attempt by social scientists to improve on intuition. Rational approaches using Shannon's information theory have argued that teams and organizations should minimize interdependence (mutual information); social psychologists have long recommended the removal of the effects of interdependence to make their data *iid* (independent, non-orthogonal, memoryless); and the social science of interdependence is disappearing. But according to experimental evidence reported by the National Academy of Sciences, the best teams maximize interdependence. Thus, navigating social reality has so far permitted only an intuitive approach to social and psychological interdependence. Absent from these conflicting paradigms is the foundation for a theory of teams that combines rationality and interdependence based on first principles which we have begun to sketch philosophically and mathematically. Without a rational mathematics of interdependence, building human-machine teams in the future will remain intuitive, based on guesswork, and inefficient.

Keywords: Interdependence · Explainable AI · Shared context · Human-machine teams

1 Introduction

Hypersonic missiles, biotechnology [35], a new nuclear national strategy [34] and human-machine teams are being introduced into battle contexts, forcing decisions to be made faster than humans can process information. In the case of the Uber car [33], it failed to alert the driver when it detected an object on its road ahead; and while the human recognized a pedestrian on the road, she reacted appropriately but too late. An unfortunate situation, but one that we free humans can learn from and fix by making the machine and the human *interdependent*, meaning that a team is able to share their view of reality, checked and confirmed by each other. This new territory with AI holds much promise. Other mistakes will follow. But when the freely governed have access to the information about their performance and mistakes [29], at the lowest level in this tragedy, we see a path to build a trust shared between machine and human as they construct shared contexts about their performance that can be verified. Human-machine teams are becoming critical to our survival, but they must be well-trained as a team, trained to abide by the rules of engagement to address ethical concerns, and then

© Springer Nature Switzerland AG 2020
J. Chen (Ed.): AHFE 2019, AISC 962, pp. 131–140, 2020.
https://doi.org/10.1007/978-3-030-20467-9_12

permitted to make their decisions autonomously. But what does it mean mathematically to be a well-trained team, and can we measure it?

For a mathematical grasp of interdependence, which works like the quantum uncertainty principle, the measurement of an interdependent object affects the object measured, indicating that action by humans and their observations of action are orthogonal, accounting for the common failure to validate social concepts with self-reports (e.g., self-esteem; implicit racial bias; personality). Theoretically, we have divided interdependence arbitrarily (the only way possible) into bistable stories (e.g., political, scientific forums); uncertainty from incomplete interpretations (e.g., the suppression of errors committed by authoritarians); and the inability to factor social states (e.g., complex legal arguments). To advance previous research success, we introduced the value of intelligence to interdependence. The prevalence in social interaction of interdependence makes social navigators rely on intelligence to amplify a team's skills with constructive interference, to mindfully use destructive interference for sharpening a team's focus, and to strategically deploy the team boundaries that block outside interference. Intelligence determines the members selected for a team (constructive); the shape of a team's structure that produces maximum entropy (MEP); and the shortest social path to overcome the obstacles to a team's superordinate goal. The quantum-like uncertainty associated with interdependence causes tension between the intuition leaders use in tradeoffs under uncertainty that shape a team and its structure to achieve MEP (e.g., the skills a team needs, their orthogonal configurations, the team's implicit communications). In this progress report, we continue to lay the groundwork for a theory of interdependence to design human-machine teams with the metrics of performance that teams need to reach their superordinate goals.

2   The Theoretical Problem

Based on our theory of interdependence, we advance knowledge by accounting for the past failure of the science of individual agents to generalize to teams. Previously, it was speculated that redundant members in a social network would improve its performance ([5], p. 716); in agreement, the National Academy of Sciences concluded that while the size of teams remained "open", it speculated that "more hands make light work" ([10], Ch. 1, p. 13). Contradicting these speculations, with theory and mathematics, we have predicted, found and replicated that redundancy impedes team performance [22, 23], an unexpected discovery that also solves the Academy's "open" problem of team size, implying that optimum team size matches problem size. Thus, mathematics for autonomous human-machine teams are critical to verify and calibrate the efficient operation of teams at MEP.

Broader Impacts: A rising concern is that AI may be an existential threat to humans. Going forward, what we have proposed [29] has the potential to benefit society by countering this threat with the establishment of limits for human-machine teams. It provides a new view of social reality based on interdependence that reinforces the value of "competing interests" (e.g., the practice of science; prosecutors and defense attorneys; politics; military actions against adversaries). Shannon's information theory and the social sciences, including economics, assume that observations of human behavior record the actual behavior that has occurred, often with self-reports of self-observed behavior [22]; e.g., if this assumption were true, alcoholic denial and deception would not exist. This assumption is unsupported by the evidence (e.g., self-reported self-esteem is unrelated to academic or work performance; in [2]), as are the results from game theory that cooperation provides for the best social good (e.g., [39]). At the heart of these rational, but false, models, interdependence is seen as a constraint (information theory; see [9]) or experimental confound (social sciences; in [21]) that must be removed to make the data *iid*. However, removing interdependence from experimental results is akin to the need to removing quantum effects at the atomic level as "pesky" [22].

While it may seem confusing, self-reported behavior and self-reported concepts often strongly correlate, but actual behavior and self-reported concepts correlate poorly [53], the latter being an example of Bohr's action-observation complementarity [22, 23]. Instead, with a theory of interdependence has come the emergence of a new phenomenon (e.g., redundancy; in [22]); a replication (e.g., redundancy; in [23]); and the value of intelligence to achieve MEP in teams, to select the best members for a team (viz., good fitness reduces a team's structural entropy), and to complete the mission of a team [27]. We have found that interdependence is the primary social resource that allows robust intelligent human teams to innovate, perform at the highest levels (MEP; in [23]), and promote social welfare by seeking an "informed assessment of competing interests" (Justice Ginsburg [17]). To achieve a model of human-machine team performance, two teams competing against each other along with a third team of neutrals are necessary to determine shared context [22]. Countering the prevailing assumptions about observation and action, individual behavior is more than a curiosity for the theory of human-machine teams because neutral individual observers enter into states of superposition (quantum-like) driven by two-sided arguments, processing information during states of interdependence to hold competing interests accountable, to gain an optimum education of the issues affecting a decision, and to best determine the decisions of teams, firms and organizations, directly applicable to human-machine teams under uncertainty.

Relevance to commerce and military: With the assumption that what applies to the US Military applies to commerce, we begin with the US Army [1] which defines joint interdependence as "the evolution of combined arms; the use of a specific military capability to multiply the effectiveness and redress the shortcomings of another." ([1], p. 3-1) The Army describes joint interdependence "as a best value solution ([1], p. 3-2);" complementary ([1], p. 3-2); to make small Army units unbeatable ([1], p. 3-9). The best scientific teams are also highly interdependent [11]; the National Academy of Sciences [10] also reported that, statistically, interdependence occurs in teams. *Two problems exist*: Cummings [11] has found that the best science teams are highly interdependent, but that interdisciplinary science teams perform the worst, and not why, which we attributed to redundancies created administratively by the National Science Foundation's promotion of interdisciplinary science teams [27]; whether a team is interdisciplinary should depend on the problem confronted; the right skills for a teammate should govern selection, based on entropy reduction reflecting fit. Thus, statistical approaches for teams must be complemented with mathematical models to select members, construct and operate human-machine teams efficiently at maximum entropy production (MEP).

Relevance to social science and society: A concern is that AI may be an existential threat to humans once autonomous AI machines understand how to replicate and extend their influence. However, building on Simon's (1989) bounded rationality, what we propose based on our research counters this threat with the establishment of limits for human-human, human-machine or machine-machine teams (see Eq. 1). Instead, we provide a mathematical view of social reality based on interdependence that reinforces the value of Ginsburg's [17] "informed assessment of competing interests" (e.g., the practice of science; prosecutors and defense attorneys; voting for a politician).

The individual model. While the individual model does not predict group effects, we shall explore team fitness based on structural entropy reduction that permits a maximum increase in MEP (see planned projects). There are significant problems with the individual model. Most social science is predicated on observations of individuals, but observations affect the performance of human workers (e.g., [41]). The social sciences, including economics, assume that observations of individual human behavior record the actual behavior that has occurred, often with self-reports of self-observed behavior [22]; e.g., if this assumption were true, alcoholic denial and deception would not exist. Another example is the science of individual decision-making which believes that forecasting is a skill that can be learned (e.g., [47]). But this model failed in its first two tests.

There is other evidence: Self-esteem fails to predict academic or work performance [2]. Validity tests indicate that the IAT failed to predict subconscious racial bias [3], and the treatments for IAT failed to show a reduction in bias [13]. The results from game theory have failed to establish that cooperation provides for the best social good [39]. In general, a meta-analysis contrasting the actions of individuals with their self-reported beliefs found weak correlations [53]. As a specific example, in a drug trial designed to protect female partners of HIV infected males (news by [8]), based on reports by the women that they complied with the drug regimen 95% of the time, the drug failed to prevent the women from contracting AIDS by their partners. Inadvertently, researchers analyzed blood levels of the women, finding that no more than 30% had anti-HIV drugs in their body at the time of the reports. Infectious disease specialist Jeanne Marrazzo said,

> There was a profound discordance between what they [the women] told us, what they brought back [empty drug vials as evidence of compliance], and what we measured [in their blood].

Based on this example and others, we conclude that the sum of the individual beliefs, attitudes & personalities of a team do not define or equal the team [22, 23].

Shannon (1948) is based on communications between individuals. Interdependence occurs when communicating across a channel, with noise but without interference. Generalizing information theory to systems, Conant [9] recommended that other than formal inter-channel communications, teams should minimize interdependence.

Social science experimenters also strive to remove the effects of interdependence [21] with the goal to increase replications by converting the data for interdependent individuals into the data of independent individuals. Two problems have occurred since; first, the failure to replicate has created a crisis in social science [32]; and, second, by throwing out the supposedly "pernicious" effects of interdependence, similar to ignoring quantum effects at the atomic level, the benefits of interdependence have been mostly ignored until the report by the National Academy of Sciences [10].

If we treat action and observation as orthogonal or independent systems in a human, that is, composed of a motor strip for action and a perceptual observer [40], Bohr's [4] action-observation model for human accounts for the problem with self-reports. We have also found that education is strongly associated with a nation's production of patents [27]; but we also found that education is not related to skills in air-to-air combat, a positive indication of orthogonality, suggesting an interdependence that needs additional study.

Interdependence. There are no parameters yet known for mathematical models of human or human-machine teams. Instead, predictions with interdependence are determined based on its factors, where interdependence is arbitrarily defined as constituted by bistability (two-sided stories; multitasking; learning); incompleteness (one-sided stories, creating uncertainty, errors); and non-factorability (untangling complex court cases; divorce in marriage or business; untangling the effects of multitasking; disentangling the endless debates searching for meaning).

Bistability: Group decision-making improves the answers to difficult questions when both sides of an issue are able to challenge each other to produce Ginsburg's "informed assessment." When a group enters an uncertain context, it needs to deploy bistable thinking to resolve the uncertainty; e.g., DOE's mismanagement of military radioactive wastes created by its one-sided decisions is being remediated with bistable debates [24]. Similar to bistability's orthogonal perspectives, multitasking between orthogonal roles in a team elicits verbal-nonverbal communications (e.g., pitcher-catcher). Based on this concept, redundancy impedes a team's performance by interfering with its communications [22, 23]. We plan to model bistability with a harmonic oscillator, with ground and elevated (emotion) states.

Incompleteness: Interdependence, especially instability when confronted by uncertainty, can reduce human error. But humans do not like challenges presented by bistable alternatives and often ignore them, increasing the likelihood of accidents (for a review, see [26]). Once machines learn how to perform as the members of a team, however, they know what is expected of the human in the loop, and if permitted by society, can step in to prevent an accident before it happens; e.g., NTSB [33] concluded that Uber car could have sent a signal warning its backup driver of the pedestrian ahead [29]. Accepting Bohr's action-observation uncertainty relation, Cohen's ([7], p. 45) signal detection theory is similar: "a narrow waveform yields a wide spectrum, and a wide waveform yields a narrow spectrum and that both the time waveform and frequency spectrum cannot be made arbitrarily small simultaneously." Incompleteness, however, means that uncertainty has become extreme.

Non-factorability: Persons in roles orthogonal to each other do not see the same social reality; e.g., husband-wife; pitcher-catcher; the members of an Army tank crew. Orthogonality leads to extended bistable debates as culpability is sorted out in legal wrangling; seeking the meaning of religion; or determining the meaning of the quantum [48]. Predicting team size is also non-factorable; theory suggests that a team must be sized to perform its function by working to solve a problem or build a product [14]; thus, unless corrupt, business mergers, military combined arms, etc., minimize redundancy when possible [23].

Groups. Group interdependence often emerges as "bewildering" ([20], p. 33). Instead of focusing on the positive effects of interdependence, group effects have become pejorative; e.g., the groupthink [19] that leads to the suppression of problem solving blamed on the Challenger shuttle disaster [18]. Suppression may promote an increase in accidents, non-innovativeness [26] or non-action; e.g., consensus-seeking rules used by minorities to control a majority [24]. Examples of possible positive effects: Mergers; teamwork; self-organization; better decisions; truth-seeking; competitive science; etc. After Jones, the study of interdependence virtually stopped, regaining credibility with the Academy's report on teams [10]. US Army [1] also considers the interdependence in teamwork to be a positive value.

Groups have been studied for over a century [43], they have figured into the founding of social psychology with the discovery of interdependence [30], but until the advent of the Science of Team Science conferences and the Academy study on teams [10], very few successes or advances have derived from the study of groups. Jones ([20], p. 33) concluded that human lives are filled with states of interdependence, but a lab study is uncontrollable. Subsequently, the study of interdependence had almost disappeared from social science until the recent study by the National Academy of Sciences [10].

Teams. From the Army [45]:

> Teamwork defined: When we refer to human-agent teaming and human-agent teamwork, we are focused on the team-level emergent properties, including states and processes, that influence performance and effectiveness (e.g., cohesion, shared mental models, shared situational awareness, coordination, and communication), rather than individual taskwork.

In response, groups and teams are different from the individuals that compose them; while the influence of a group may be greater than sum of individual contributions [30], the power of a well-functioning team emerges a team's structure which allows its members to fully cohere to its goal [10], implying reduced degrees of freedom (*dof*); e.g., an agent succeeds when its independent taskwork becomes coordinated with the team's work by reducing the team's structural entropy as individual identity disappears into team membership, allowing MEP for a team's goal [23]. Supporting Von Neumann's subadditivity in Eq. 1–2, if a team's structure becomes stable and functional (like a crystal), the entropy goes to zero:

$$lim_{structure \to 1} \ln(dof) = 0 \qquad (1)$$

Based on Eq. (1), that gives us subadditivity:

$$S(x, y) \leq S(x) + S(y). \qquad (2)$$

Teams with highly-functioning interdependent members outperform the same set of individuals who compose the team [10]; well-performing teams do not use bistability until a team's plan goes awry and a new course of action has to be decided. Teamwork, its dedicated function, makes highly interdependent teams behave differently than groups. For that reason, predictions about teams can be modeled mathematically to complement statistical methods. However, replacing a well-trained member with a new member causes the performance of a team to deteriorate initially [10], indicating that the structure is under stress, suggesting that non-interdependent groups obey Shannon; i.e., in Eq. 3, as members become independent:

$$H(x, y) \geq H(x), H(y) \tag{3}$$

Equation (3) indicates too much Shannon information is generated by a failing team (e.g., divorce); Eq. (1) indicates that too little information is generated by successful teams.

From Strong ([45], p. 7),

> specific long-term goal to provide the fundamental research necessary to develop individualized, adaptive technologies that promote effective teamwork in novel groups of humans and intelligent agents ... focused on the team-level emergent properties ...

In reply, while experimental approaches focused on individuals has not led to satisfactory theory of teams, based on interdependence theory, by freely choosing candidates for a position on a team in a state of dynamic performance as during a competition, or by rejecting those who do not fit with a team, where structural fitness is measured with a drop in entropy, "Individualized approaches" can account for emergent process in teams and its performance [23].

Future plans to advance research. Theory and mathematics must continue to be developed to model the orthogonal effects of complementarity (Eq. 1). Theory needs to be developed for the structure of teams, the value and management of team boundaries [14, 22], and the shaping of MEP. Team structure when stable minimizes energy consumption, affords a separation from others to reduce interference, and provides minimal defense, but structures require maintenance. We have adopted Prigogine's [38] idea of low entropy production (LEP) for a structure; given a struggle for sources of energy, with only so much available, a structure with least LEP permits MEP to be directed at a team's primary work. Take as an extreme example, divorce (business, like the turmoil between Nissan and Renault; the UK's Brexit; divorce for a couple) occurs when a structure is being dismantled, leaving little energy available to accomplish a team's mission. Conflict may also arise when a team's structure is inappropriate for the team (e.g., if a team's configuration does not address its problem sufficiently), or from poor individual fit [10]. With an agent-based model, we plan to develop an Ising or Bethe lattice model of a team's boundaries and internal conflict.

We have predicted that a member's fitness is similar to a crystal (or molecule) before and after a crystal has been constructed. We anticipate, similar to a crystal, that when a team's configuration is correct and the last member entering the team is a good fit for a team's structure, structural entropy reduces [22]. Theory also needs to be developed for the work by a team; if a team's function has a work output or product, the structure of the team and the members chosen must be adjusted to maximize MEP [14]. But by giving a minority control of its agenda, consensus-seeking groups are often unable to recommend concrete actions, producing poor output (e.g., the European Union's white paper about being held hostage by minority control; in [52], p. 29). Theory also needs to development for the control of a team. Should teams be regimented (where they can be suppressed, like humans); or should they be given rules of engagement (ROE), trained until they are verifiably able to produce MEP to accomplish their mission, and allowed to perform autonomously, a hypothesis to be tested.

Technology, measuring entropy, can provide metrics of performance for the team to observe itself. It can also determine when a team member is contributing or not to a team.

3 Conclusions

Despite success over past decades with AI [36], and now with machine learning (ML), AI for many new technologies like self-driving cars, remains unable to explain itself [37]. Instantiated in machines performing delicate, high-risk tasks of increasing importance to military success and social well-being, including with nuclear weapons [34] and hypersonic missile defense [35], as these tasks operate at speeds too fast for humans to process, technology is required to explain context to build a "shared mental model" for a machine's interpretation of reality that builds shared trust when its performance can be verified [29]. Human-machine teams are becoming critical to our survival, but they must be well-trained with verifiable performance, trained to abide by the rules of engagement, before allowing them to make their decisions autonomously.

References

1. ADP-1. The Army. 2012. From https://www.army.mil/e2/c/downloads/303969.pdf
2. Baumeister, R.F., Campbell, J., Krueger, J., and K. Vohs.: Exploding the self-esteem myth, Sci. Am., **292**(1), 84–91 (2005 Jan). https://www.nature.com/scientificamerican/journal/v292/n1/full/scientificamerican0105-84.html
3. Blanton, H., Klick, J., Mitchell, G., Jaccard, J., Mellers, B., Tetlock, P.: Strong claims and weak evidence: Reassessing the predictive validity of the IAT. J. Appl. Psychol. **94**(3), 567–582 (2009)
4. Bohr, N.: Science and the unity of knowledge. In: Leary, L. (ed.) The unity of knowledge, pp. 44–62. Doubleday (1955)
5. Centola, D., Macy, M.: Complex contagions and the weakness of long ties. Am. J. Sociol. **113**(3), 702–734 (2007). https://doi.org/10.1086/521848
6. Clarke, E.M., Grumberg, O., Peled, D.A.: Model checking. MIT Press, Cambridge (1999)
7. Cohen, L.: Time-frequency analysis: theory and applications. Prentice Hall Signal Processing Series (1995)
8. Cohen, J.: Human nature sinks HIV prevention trial. Science **351**, 1160 (2013). https://www.sciencemag.org/news/2013/03/human-nature-sinks-hiv-prevention-trial
9. Conant, R.C.: Laws of information which govern systems. IEEE Trans. Sys. Man Cybern. **6**, 240–255 (1976)
10. Cooke, N., Hilton, M. E.: Enhancing the effectiveness of team science. National Research Council, National Academies Press, Washington (DC) (2015)
11. Cummings, J.: Team science successes and challenges. In: NSF Workshop Fundamentals of Team Science and the Science of Team Science, Bethesda MD (June 2, 2015). https://www.ohsu.edu/xd/education/schools/school-of-medicine/departments/clinicaldepartments/radiation-medicine/upload/12-_cummings_talk.pdf
12. Ellis, J.: American dialogue. The founders and us. Knopf (2018)

13. Emerson, J.: Don't give up on unconscious bias training—Make it better. Harv. Bus. Rev. (4/28/2017). https://hbr.org/2017/04/dont-give-up-on-unconscious-biastraining-make-it-better

14. England, J.: Statistical physics of self-replication. J. Chem. Phys. **139**(121923) (2013)

15. Flint, J., Hagey, K.: CBS ups stakes in feud with Redstones. The media company turns to 'nuclear option' to eliminate Redstones' voting control and block a Viacom merger. Wall Street J (5/14/2018) https://www.wsj.com/articles/cbs-goes-on-attack-againstredstones-suing-controlling-shareholder-for-breaching-fiduciary-duty-1526308237

16. Folley, A.: Louisiana law repealed when voters helped endeda Jim-Crow era law that has long influenced Louisiana's legal system. NOLA.com reported (11/7/2018)

17. Ginsburg, R.: American Electric Power Co., Inc., et al. v. Connecticut et al. 564 US 410 Supreme Court (2011). https://supreme.justia.com/cases/federal/us/564/410

18. Gouran, D.S., Hirokawa, R.Y., Martz, A.E.: A critical analysis of factors related to decisional processes involved in the Challenger disaster, Comm. Stud. **37**(3), 118–135 (1986, September). https://doi.org/10.1080/10510978609368212

19. Janis, I.L.: Groupthink. Houghton, New York (1971)

20. Jones, E.: Major developments in five decades of social psychology. In: Gilbert, D. T., Fiske, S.T., Lindzey, G. (eds.) The Handbook of Social Psychology, vol 1. McGraw-Hill, 1998 (see p. 33)

21. Kenny, D. A., Kashy, D. A., Bolger, N.: Data analysis in social psychology. In: Gilbert, D.T., Fiske, S.T., Lindzey, G. (eds.) The Handbook of Social Psychology (pp. 233–265). New York, NY, US: McGraw-Hill (1998)

22. Lawless, W.F.: The entangled nature of interdependence. bistability, irreproducibility and uncertainty. J. Math. Psy. **78**, 51–64 (2017a)

23. Lawless, W.F.: The physics of teams: Interdependence, measurable entropy and computational emotion. Front. Phys. **5**(30) (2017b)

24. Lawless, W.F., Akiyoshi, M., Angjellari-Dajcic, Fiorentina, Whitton, J.: Public consent for the geologic disposal of highly radioactive wastes and spent nuclear fuel. Int. J. Environ. Stud. **71**(1), 41–62 (2014)

25. Lawless, W.F., Sofge, D.: The intersection of robust intelligence and trust in autonomous systems, Chapter 12. In: Mittu, R., Sofge, D., Wagner, A., Lawless, W.F., The intersection of robust intelligence and trust: Hybrid teams, firms and systems. Springer (2016)

26. Lawless, W.F., Mittu, R., Sofge, D., Russell, S. (eds.): Autonomy and artificial intelligence: a threat or savior? Springer, New York (2017)

27. Lawless, W.F.: (under review) Interdependence, shared context and uncertainty for human-machine teams. Towards the mathematics of explainable AI, Frontiers of Science

28. Lawless, W.F., Mittu, R., Sofge, D., Moskowitz, I.S., Russell, S. (eds.): Artificial intelligence for the internet of everything. Elsevier (2019a, in press)

29. Lawless, W.F., Mittu, R., Sofge, D.A., Hiatt, L.: Introduction to the Special Issue, "Artificial intelligence (AI), autonomy and human-machine teams: Interdependence, context and explainable AI," AI Magazine (2019b, in press)

30. Lewin, K.: Field theory of social science. Selected theoretical papers, D. Cartwright (Ed.). Harper and Brothers (1951)

31. Millen, R.: "Cultivating strategic thinking: the Eisenhower model," Parameters (2012, Summer), https://ssi.armywarcollege.edu/pubs/parameters/articles/ 2012summer/Millen.pdf

32. Nosek, B.: Estimating the reproducibility of psychological science. Science, **349**(6251), 943 (2015). http://science.sciencemag.org/content/349/6251/aac4716

33. NTSB.: Preliminary report released for crash involving pedestrian, Uber Technologies, Inc., test vehicle. Technical report, National Transportation Safety Board, NTSB (24/5/2018) https://www.ntsb.gov/news/press- releases/Pages/NR20180524.aspx, 2018

34. OSD.: Nuclear Posture Review, Office of the Secretary of Defense (OSD) (2018a) https://media.defense.gov/2018/Feb/02/2001872886/-1/-1/1/2018-NUCLEARPOSTURE-REVIEW-FINAL-REPORT.PDF
35. OSD.: National Defense Strategy, Office of the Secretary of Defense (OSD) (2018b) https://dod.defense.gov/Portals/1/Documents/pubs/2018-National-Defense-Strategy-Summary.pdf
36. Pearl, J.: Reasoning with cause and effect. AI Magazine, 23(1), 95–111, 2002 http://ftp.cs.ucla.edu/pub/stat_ser/r265-ai-mag.pdf
37. Pearl, J., D. Mackenzie.: AI can't reason why. The current data-crunching approach to machine learning misses an essential element of human intelligence. Wall Str. J. (2018). https://www.wsj.com/articles/ai-cant-reason-why-1526657442
38. Prigogine, I.: Time, structure and fluctuations, Nobel Lecture, pp. 263–285 (12/8/1977)
39. Rand, D., Nowak, M.: Human cooperation. Cognitive Sciences, 17(8), 413–425 (2013). https://static1.squarespace.com/static/51ed234ae4b0867e2385d879/t/54ac6169e4b00f7c5fc76aa2/1420583312697/human-cooperation.pdf
40. Rees, G., Frackowiak, R., Frith, C.: Two modulatory effects of attention that mediate object categorization in human cortex. Science 275, 835–838 (1997)
41. Roethlisberger, F., Dickson, W.: Management and the worker. Cambridge University Press (1939)
42. Shannon, C.: A mathematical theory of communication. Bell Syst. Tech. J. 27, 379–423, 623–656 (1948)
43. Simmel, G.: The persistence of social groups. AJS 3, 662–698 (1897)
44. Simon, H.: Bounded rationality and organizational learning. Tech. Rep. AIP 107, CMU, Pittsburgh, PA (23/9/1989)
45. Strong, USArmy: Strengthening teamwork for robust operations in novel groups (2018)
46. Taplin, H.: Can China's red capital really innovate? Wall Str. J. (2018). https://www.wsj.com/articles/can-chinas-red-capital-really-innovate-1526299173
47. Tetlock, P., Gardner, D.: Superforecasting: the art and science of prediction. Crown (2015)
48. Weinberg, S.: The trouble with quantum mechanics. The New York Review of Books http://www.nybooks.com (2017)
49. Weyers, B., Bowen, J., Dix, A., Palanque, P. (eds.): The handbook of formal methods in human-computer interaction. Gewerbestrasse: Springer International (2017)
50. Wilkes, W.: How the world's biggest companies are fine-tuning the robot revolution. Automation is leading to job growth in certain industries where machines take on repetitive tasks, freeing humans for more creative duties. Wall Str. J. (14/5/2018). https://www.wsj.com/articles/how-the-worlds-biggest-companies-are-finetuning-the-robot-revolution-1526307839
51. Wing J.M.: A specifier's introduction to formal methods. Computer 23(9), 8, 10–22, 24 (1990)
52. WP: White Paper. European governance (COM (2001) 428 final; Brussels, 25.7.2001). Brussels, Commission of the European Community (2001)
53. Zell, E., Krizan, Z.: Do people have insight into their abilities? A metasynthesis? Perspect. Psychol. Sci. 9(2), 111–125 (2014)

Neurophysiological Closed-Loop Control for Competitive Multi-brain Robot Interaction

Bryan Hernandez-Cuevas[1](✉), Elijah Sawyers[1], Landon Bentley[1],
Chris Crawford[1], and Marvin Andujar[2]

[1] The University of Alabama, Tuscaloosa, AL 35401, USA
{byhernandez, esawyers, lcbentley}@crimson.ua.edu,
crawford@cs.ua.edu
[2] University of South Florida, Tampa, FL 33620, USA
andujar1@usf.edu

Abstract. This paper discusses a general architecture for multi-party electroencephalography (EEG)-based robot interactions. We explored a system with multiple passive Brain-Computer Interfaces (BCIs) which influence the mechanical behavior of competing mobile robots. Although EEG-based robot control has been previously examined, previous investigations mainly focused on medical applications. Consequently, there is limited work for hybrid control systems that support multi-party, social BCI. Research on multi-user environments, such as gaming, have been conducted to discover challenges for non-medical BCI-based control systems. The presented work aims to provide an architectural model that uses passive BCIs in a social setting including mobile robots. Such structure is comprised of robotic devices able to act intelligently using vision sensors, while receiving and processing EEG data from multiple users. This paper describes the combination of vision sensors, neurophysiological sensors, and modern web technologies to expand knowledge regarding the design of social BCI applications that leverage physical systems.

Keywords: BCI · EEG · Competitive · Architecture · Robotics · Multi-party · Drones · Intelligent

1 Introduction

Several EEG-based brain-controlled robot architectures have been previously presented. Much of these prior investigations involve medical applications such as brain-controlled wheelchairs [5, 9, 13] and neuroprosthetics [19]. While these applications are highly relevant, researchers have also begun exploring non-medical BCI applications. Examples of this approach include drowsiness detection while driving [14], workload estimation [7, 11], gaming [6], engagement monitoring [28] neurofeedback training [28], and neuromarketing [2]. Many of these applications use passive BCIs which often aim to enrich human-computer interaction with implicit information derived from brain activity [30, 31]. Furthermore, previous discussions involving adaptive systems that leverage neurophysiological information mostly focus on single user interactions.

© Springer Nature Switzerland AG 2020
J. Chen (Ed.): AHFE 2019, AISC 962, pp. 141–149, 2020.
https://doi.org/10.1007/978-3-030-20467-9_13

Conventional input modalities (e.g., mouse and keyboard) are expected to out-perform BCI-based control. However, researchers have recently explored various opportunities to turn these shortcomings into challenges that engage users while interacting with BCI applications [20, 25]. This approach has mainly been investigated in the context of gaming. Although games commonly involve a social component, research regarding social brain-computer interfacing is limited. Previous investigations on the social dynamics of multi-party BCI applications usually involved multiple users interacting with virtual objects [12, 23].

A select few social brain-computer interfacing applications feature physical objects. However, these objects do not have the ability to sense or react to the surrounding environment. Consequently, these systems are limited to continuous manual control and prone to errors in dynamic environments (e.g., motion of air). Both of these drawbacks could lead to challenges as the desired behavior of the systems become more complex (e.g., linear movement vs. linear movement while avoiding obstacles). While previous work provides important first steps, there is not much knowledge concerning ways to leverage control architectures to support social BCI experiences with intelligent physical systems such as robots. In response, this work aims to expand knowledge on the design of control systems that support social BCI applications featuring robots.

2 Background

Current research work in individual and multi-party passive BCI applications mostly targets clinical environments with serious restrictions to other possible integrations. However, there are others that consider multi-party brain-computer interfaces since it allows performance increase in group activities. Also, these projects involving physical devices are primarily focused on controlling mobile robots.

BCI applications have mostly targeted clinical environments. EEG-based control signals for moving physical devices have been researched in that domain. An example is a project where electroencephalography signals from an individual would move an electric wheelchair toward two basic directions (right or left). A recursive algorithm recognizes signals from the brain and translates input into controls for the wheelchair [29].

Currently, there is abundant work for BCI in mobile robot controls. One example is the study conducted to discover possible applications for EEG-based input systems to control mobile robots. Here, while fine control was not achieved, initial hands-free operation of small mobile robotic devices was attained [1]. Various other researchers explore distinct applications for the control of these robots; examples include evaluation systems for real-time operations [22], EEG-based control of telepresence mobile robots [8] and adaptive user interfaces with multiple degrees-of-freedom in robotic controls for more effective information transfer [10].

Furthermore, work has been done in neuroprosthetic devices. Research involves the use of BCIs based on steady-state visual evoked potentials to provide people the ability to control an electric hand prosthesis using their brain [18]. Similar work has also been completed on developing more natural, intuitive neuroprosthesis control using non-invasive EEG-based BCIs. The goal is to improve the quality of life for people with cervical spinal cord injuries [17]. However, all the aforementioned work is directly related to clinical applications alone and does not explore multi-party, social BCI integration.

As discussed in [25], BCI has been used for control in social and gaming environments. However, most BCI applications leverage traditional visual approaches as feedback. Recently additional research has reviewed multi-brain games based on multiparty brain-computer interfacing. They investigate ways of merging EEG activity into a social context for collaboration and competition [24]. More recent research has discussed how multi-brain BCI can become a paradigm for artistic and gaming applications, where the information from EEG activity becomes valuable. Measurements of electroencephalography data from different users are needed to join into a single system that translates information into controls for multiple physical devices [21].

In this paper, we leverage the use of passive BCIs in multi-party, competitive environments. Our goal is to explore a hybrid architecture for multiple neurophysiological interactions in a competitive social context to control aerial robotic devices. In the following sections we discuss a hybrid architectural model designed for a competitive activity with multiple brain-computer interfacing to control drones that intelligently follow a line to avoid possible hazardous situations.

3   Architectural Design

The design for this model is based on the principle of subsumption architecture [4]. The system contains layers (or modules) with individual purposes. A central module - here called *central controller* - receives multiple inputs from the connected EEG devices. Data from the devices is sent to the controller and forwarded to the multiple robots in the network as commands for action. While the robots complete their task, they sense their environment and return raw data. When the central controller receives that data from the robots, it is sent to the error calculation module for re-adjusting the parameters that provide necessary reactions to the sensed stimuli. The process is continuously occurring, with interactions between the central controller and the data from each device comprising the closed-loop controller. This way, robots are intelligently reacting to their surroundings. The following subsections detail the way each component works, and the modules associated to each one.

3.1   Central Controller

The central controller is the core of our proposed model. Each EEG device and robot that connects to the network will directly communicate through this controller. More than only being an intermediary component, it contains various modules that every connected device will use. For example, if the purpose is to maintain a robot from colliding with obstacles, a module to calculate the distance to the object will be included in the central controller.

3.2   EEG Device

Robots are controlled through BCI from competing users. This means there will be multiple EEG devices connecting to the same central controller, each transmitting raw data that should be processed and analyzed. This shows that modules for these interactions must be implemented. Minimum required layers are for EEG data reception and conversion into robot commands. Those commands would be used in another module that directly communicates with the robots.

3.3   Robot

Robots connected to the network should contain their own set of modules. The purpose of each robot is to receive commands from the central controller, sense environmental stimulus and send back raw data on sensory information. In this case, the central controller must have a minimum of three modules for robot interaction: command sender, raw data receiver and data manager. The data manager can represent multiple modules, since the received information could need diverse calculations before being sent back to the robot. In general, the data collection and calculations produce intelligent adjustments for the robot; meaning it can react to given possible situations. These interactions between the robot and the central controller represent the closed-loop system.

4   Model Implementation

In this paper we will focus on a specific case where the proposed architecture can be applied. The example is composed of BCI interaction with two EEG devices, two drones, and a central controller that resides in a ROS [27] instance. This use case is described as follows: two users send velocity commands (via regulation of brain activity) to each of their respective drones, while the closed-loop controller provides the drones intelligent reactions to follow a line without detouring.

The main purpose for the drones will be to always follow a line by intelligently reacting to any deviation. For the human subjects, the goal is to maintain focus, which is translated into a velocity for each drone.

First, we must evaluate which modules are needed in this example situation. For EEG devices, the data must be received and translated into velocity commands. Therefore, there are two modules needed for these specific devices. The receiver module gets the raw EEG data sent by the EEG device. Then, this information is passed to a conversion module where data preprocessing and classification will be made to create a velocity command. Once this is implemented, the next step is to send the command to the drone.

Each drone receives velocity commands from a user. To provide the robot with a specific task, in this case 'follow a line', additional core modules are required: data receiver, sender, and line error calculator. Figure 1 provides an illustration of the model. The first two modules are basic layers for receiving the EEG-translated command and sending commands to the drone, which are comprised of subscribers and publishers in ROS. The third layer, the line error calculator, will be used to receive direct raw image data from the drone and calculate the error between the line to be followed and the current path. Using the calculated error, this information will be piped into the drone sender module. The system would then output intelligent reactions to the drone. Next, the functionality for the closed-loop control will be discussed in detail.

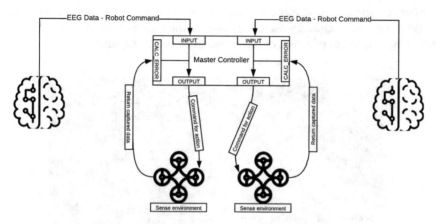

Fig. 1. Architectural model for intelligent multi-brain robot interaction

5 Closed-Loop Control Intelligence

The closed-loop control in this model proposition is located in the interactions between robotic devices and the master controller. First, the design process must begin by recognizing the layers involved: sender, receiver, and error calculator. Since the subscriber (receiver) and publisher (sender) are both ROS-specific methods, we shall focus on line error calculations. During this study, our team worked with Parrot Bebop 2 drones [26]. Also, Python and OpenCV [3] were utilized in the ROS ecosystem.

Our approach for calculating line error starts with the compilation of raw image data. The drone captures real-time images of its current position, which is then processed to calculate an error between the desired path and current position. This goal can be achieved with image capture, line detection, and linear velocity adjustments.

5.1 Image Capture

The drone being utilized in this pilot implementation contains a frontal camera with angular inclination capabilities. This is necessary to capture the current path and decide if there is a line to follow. For this particular experiment, the task is to follow a red line on the floor. The Parrot Bebop 2 drone captures images and streams them in the network to subscribers in the system. After images are intercepted through the respective protocol, they are piped into the error calculation module. This module takes an image as input, detects a line, and then returns the calculated error. The image capture protocol used in this experiment is the bebop autonomy [16] driver for ROS.

Through this protocol, captured images can be directly streamed to the error calculation module through ROS callback methods.

5.2 Line Detection

After getting the image, it is processed to find a red line, as seen in Fig. 2, and provide possible adjustments to the robot. For this, the image stream is converted into the HLS format and a filter is applied to isolate red pixels. The median x-value of all red pixels is then compared to the midpoint of the image to calculate the error.

Fig. 2. Drone track design with red lines during Brain-Drone Race 2016.

The goal of detecting the line is the error calculation. When the drone recognizes the presence of a line, it should be able to move to that exact location. Thus, the aforementioned algorithm concludes by returning one of three possible values: 1, −1 or 0. Whenever the error value is 1, it represents the line being located to the right of the drone. If the error is −1, the line is located to the left, and if it is 0 then the line is in a small neighborhood of the midpoint. This can be updated for different specific cases, depending on the desired accuracy. Each error value will trigger a linear-y change on the drone permitting a sideways movement and stopping whenever the goal of having zero error is fulfilled.

5.3 Linear Velocity Adjustments

Finally, the first iteration on the closed-loop control is finished when the linear-y value updates accordingly. The error value determines one of three possibilities: positive, negative or zero velocity. In this example implementation there is always a constant velocity, with changes only in the direction when necessary. For example, if the error result is positive, the drone should be given a negative velocity. After the necessary adjustments are sent to the robot, the process repeats. Linear-x velocity is manipulated in response to commands derived from data captured from the EEG devices. In the case of an active BCI application the EEG data is translated into forward commands that response to motor imagery task. In this case, the drone receives a forward command when a user imagines a motor task (e.g. closing right or left hand). In a passive BCI application linear-x velocity commands can be derived as a function of band powers (see [15] for a review of classification algorithms).

6 Conclusion

This work presents an architecture that integrates multiple drones and BCI devices in a social setting while providing the physical objects an ability to sense and react to external stimuli. During multi-party activities, the robot should be able to adjust itself in the case that any unwanted action is being applied to it (e.g., wind drafts). Therefore, a solution is to make each robot stream its output to a central module that processes errors and returns a solution in the form of an action for the device to undertake. This type of closed-loop control system can provide accurate and fast data through the physical device's sensors.

Through the use of this architecture, a multi-party and social brain-computer interface environment can be successfully created. The physical device will become intelligent and self-aware to take decisions whenever outside stimuli are involved in the process, not only providing more stability but also safety and robustness. As non-medical applications of BCI continue emerging, the social aspect will evolve into more complex fields including better devices and more advanced competitive activities.

7 Future Work

Going forward, more algorithms will be analyzed and created to improve the current system. Image processing can become inaccurate whenever lighting and other external causes start to become more prominent. Therefore, one possible area of research is to advance the current image processing technologies by adding better light filtering, color manipulation and error calculations. Moreover, the use of machine learning methods is a primary target to explore for future implementations of this architecture. Through the application of these types of techniques, the robot can become more aware of its surroundings as it learns from past experiences and corrects itself in a more efficient manner.

References

1. Amai, W., Fahrenholtz, J., Leger, C.: Hands-free operation of a small mobile robot. Auton. Robot. **11**(1), 69–76 (2001). https://doi.org/10.1023/A:1011260229560
2. Ariely, D., Berns, G.S.: Neuromarketing: the hope and hype of neuroimaging in business. Nat. Rev. Neurosci. **11**(4), 284–292 (2010)

3. Bradski, G., Kaehler, A.: OpenCV. Dr. Dobbs journal of software tools 3 (2000)
4. Brooks, R.A.: A robust layered control system for a mobile robot. IEEE J. Robot. Autom. (1986). https://doi.org/10.1109/JRA.1986.1087032
5. Carlson, T., del R. Millan, J.: Brain-controlled wheelchairs: a robotic architecture. IEEE Robot. Autom. Mag. **20**(1), 65–73 (2013)
6. Chanel, G., Rebetez, C., Btrancourt, M., Pun, T.: Emotion assessment from physiological signals for adaptation of game difficulty. IEEE Trans. Syst. Man Cybern. Part A: Syst. Hum. **41**(6), 1052–1063 (2011)
7. Cutrell, E., Tan, D.: BCI for passive input in HCI. In: Proceedings of CHI, vol. 8, pp. 1–3. Citeseer (2008)
8. Escolano, C., Antelis, J.M., Minguez, J.: A telepresence mobile robot controlled with a noninvasive brain-computer interface. IEEE Trans. Syst. Man Cybern. Part B: Cybern. **42** (3), 793–804 (2012). https://doi.org/10.1109/TSMCB.2011.2177968
9. Galn, F., Nuttin, M., Lew, E., Ferrez, P.W., Vanacker, G., Philips, J., del R. Milln, J.: A brain-actuated wheelchair: asynchronous and non-invasive braincomputer interfaces for continuous control of robots. Clin. Neurophysiol. **119**(9), 2159–2169 (2008)
10. Gandhi, V., Prasad, G., Coyle, D., Behera, L., McGinnity, T.M.: EEG-based mobile robot control through an adaptive brain-robot interface. IEEE Trans. Syst., Man, Cybern.: Syst. **44** (9), 1278–1285 (2014). https://doi.org/10.1109/TSMC.2014.2313317, http://ieeexplore.ieee. org/lpdocs/epic03/wrapper.htm?arnumber=6787110
11. Grimes, D., Tan, D.S., Hudson, S.E., Shenoy, P., Rao, R.P.: Feasibility and pragmatics of classifying working memory load with an electroencephalograph. In: Proceedings of the SIGCHI Conference on Human Factors in Computing Systems, pp. 835–844. ACM (2008)
12. Hjelm, S.I., Browall, C.: Brainball-using brain activity for cool competition. In: Proceedings of NordiCHI, vol. 7 (2000)
13. Iturrate, I., Antelis, J.M., Kubler, A., Minguez, J.: A noninvasive brain-actuated wheelchair based on a p300 neurophysiological protocol and automated navigation. IEEE Trans. Rob. **25**(3), 614–627 (2009)
14. Lin, C.T., Chang, C.J., Lin, B.S., Hung, S.H., Chao, C.F., Wang, I.J.: A real-time wireless braincomputer interface system for drowsiness detection. IEEE Trans. Biomed. Circuits Syst. **4**(4), 214–222 (2010)
15. Lotte, F., Congedo, M., Le´cuyer, A., Lamarche, F., Arnaldi, B.: A review of classification algorithms for eeg-based brain–computer interfaces. J. Neural Eng. **4**(2), R1 (2007)
16. Monajjemi, M.: bebop_autonomy. http://wiki.ros.org/bebopautonomy
17. Müller-putz, G.R., Pereira, J., Ofner, P., Schwarz, A., Dias, C.L., Kobler, R.J., Hehenberger, L., Pinegger, A., Sburlea, A.I.: Towards non-invasive brain-computer interface for hand/arm control in users with spinal cord injury. In: 2018 6th International Conference on Brain-Computer Interface (BCI), pp. 1–4 (2018)
18. Müller-Putz, G.R., Pfurtscheller, G.: Control of an electrical prosthesis with an SSVEP-based BCI. IEEE Trans. Biomed. Eng. **55**(1), 361–364 (2008). https://doi.org/10.1109/TBME.2007.897815
19. Müller-Putz, G.R., Scherer, R., Pfurtscheller, G., Rupp, R.: Eeg-based neuroprosthesis control: a step towards clinical practice. Neurosci. Lett. **382**(1), 169–174 (2005)
20. Nacke, L.E., Kalyn, M., Lough, C., Mandryk, R.L.: Biofeedback game design: using direct and indirect physiological control to enhance game interaction. In: Proceedings of the SIGCHI Conference on Human Factors in Computing Systems, pp. 103–112. ACM (2011)
21. Nijholt, A.: Multi-modal and multi-brain-computer interfaces: a review. In: 2015 10th International Conference on Information, Communications and Signal Processing, ICICS 2015 (2016). https://doi.org/10.1109/ICICS.2015.7459835

22. Nijholt, A., Allison, B.Z., Jacob, R.J.K.: Brain-computer interaction: can multimodality help? In: Proceeding ICMI '11 Proceedings of the 13th International Conference on Multimodal Interfaces, pp. 35–39 (2011). https://doi.org/10.1145/2070481.2070490

23. Nijholt, A., Gürkök, H.: Multi-brain games: cooperation and competition. In: International Conference on Universal Access in Human-Computer Interaction, pp. 652–661. Springer (2013)

24. Nijholt, A., Gürkök, H.: Multi-brain games: cooperation and competition. Lecture Notes in Computer Science (including subseries Lecture Notes in Artificial Intelligence and Lecture Notes in Bioinformatics) 8009 LNCS (PART 1), pp. 652–661 (2013). https://doi.org/10.1007/978-3-642-39188-0-70

25. Nijholt, A., Reuderink, B., Bos, D.O.: Turning shortcomings into challenges: Brain-computer interfaces for games. Lecture Notes of the Institute for Computer Sciences, Social-Informatics and Telecommunications Engineering, vol. 9, LNICST (2), pp. 153–168 (2009). https://doi.org/10.1007/978-3-642-02315-615, http://dx.doi.org/10.1016/j.entcom.2009.09.007

26. Parrot, S.: Parrot bebop 2 (2016). Retrieved from Parrot.com: http://www.parrot.com/products/bebop2

27. Quigley, M., Conley, K., Gerkey, B., Faust, J., Foote, T., Leibs, J., Wheeler, R., Ng, A.Y.: Ros: an open-source robot operating system. In: ICRA Workshop on Open Source Software, vol. 3, p. 5. Kobe, Japan (2009)

28. Szafir, D., Mutlu, B.: Pay attention!: designing adaptive agents that monitor and improve user engagement. In: Proceedings of the SIGCHI Conference on Human Factors in Computing Systems, pp. 11–20. ACM (2012)

29. Tanaka, K., Matsunaga, K., Wang, H.O.: Electroencephalogram-based control of an electric wheelchair. IEEE Trans. Rob. 21(4), 762–766 (2005). https://doi.org/10.1109/TRO.2004.842350

30. Zander, T.O., Kothe, C.: Towards passive brain–computer interfaces: applying brain–computer interface technology to human–machine systems in general. J. Neural Eng. 8(2), 025005 (2011)

31. Zander, T.O., Kothe, C., Jatzev, S., Gaertner, M.: Enhancing human-computer interaction with input from active and passive brain-computer interfaces. In: Brain-Computer Interfaces, pp. 181–199. Springer (2010)

Robotic Monocular SLAM in Wastewater Treatment Plants with a Sampling Device

Edmundo Guerra[1], Yolanda Bolea[1], Rodrigo Munguia[2], and Antoni Grau[1(✉)]

[1] Automatic Control Department, BarcelonaTech, 08034 Barcelona, Spain
{edmundo.guerra,yolanda.bolea,antoni.grau}@upc.edu
[2] Department of Computer Science, CUCEI,
University of Guadalajara, 44430 Guadalajara, Mexico
rodrigo.munguia@upc.edu

Abstract. A novel application is presented in this paper. Wastewater treatment plants need to sample water to follow the cleaned process continuously. This task is very tedious for laboratory workers and, mainly, can be done in reduced areas of the large basins that form the whole plant, specifically only in the edges of those basins. The new proposal is to enlarge the sampling area and reduce the load for workers. A new unmanned aerial vehicle has been designed together with a novel tool to take water samples. Moreover, a new architecture for the mission plan is presented in form of multi-agents. Experiments have been done in order to test the feasibility of the proposal.

Keywords: Monocular SLAM · HRI · Mobile robotics ·
Wastewater treatment plant

1 Introduction

Management and supervision techniques for industrial scale processes have historically been driven by the changes produced by technical innovations. With each increase in efficiency and scale due improvements in the technologies used to produce the good and services, an increased production had to be managed. In these production tasks the automation of processes has made the human work (as physical effort) largely redundant in the heavier tasks. For more complex tasks human presence is still required, given the enormous versatility of the human cognitive abilities when compared with automated systems, which must be built according to precise specifications, and require great efforts to implement any change. This kind of tasks is especially prevalent in supervision, monitoring and management functions, where the expertise provided by human elements in a system is extremely difficult to model in a general way. In this context, the best results achieved are specialized support tools 1, which routinely can beat human in specific well-delimited tasks, but lack generality and flexibility to replace a human.

These monitoring and supervision tasks can present repetitive operations which could be subject to automation thanks to the increased pace of autonomous robotics

© Springer Nature Switzerland AG 2020
J. Chen (Ed.): AHFE 2019, AISC 962, pp. 150–162, 2020.
https://doi.org/10.1007/978-3-030-20467-9_14

development observed in the last decades. The last revolutions in the field of autonomous robotics is coming in the form of unmanned aerial vehicles (UAV) which, thanks to developments in battery technology, microelectromechanical systems (MEMS), and sensor technologies including IMU (inertial measurement unit), GPS, RFs, have gained a presence in several industries, like specialized logistic services, audio-visual production, construction surveying, etc.… Still, in this industries the UAVs operate always under human direct control, or at least supervision, in most cases, with limited autonomous applications, and only in the surveying field are gaining a significant foothold [1–4]. Applications in this industry often benefit greatly from their mobility and freedom of movement, an in many cases, supposes a reduction in the risks assumed by human workers.

In order to automatize these operations, the robotic UAVs have to have autonomy of operation enough to act on their own agency: they will have to take decisions with respect a known task and environment that are still dynamic and cannot be completely modelled and represent. There are several design approaches [5] and technologies [6] when trying to produce robots with these capabilities, being one of the most fruitful the multi-agent system (MAS). A good example of how to exploit all these features in a robotics MAS can be found in [7], where based on MAS methodologies [8], a solution for a surveying surface vehicle is implemented over the ROS framework [9].

With respect to the problem of sampling and autonomous robots, in previous works [10] authors presented a general architecture for a single unmanned aerial vehicle working jointly with a unmanned surface vehicle in a wastewater plant. This work sets the basis of this new research, where a UAV is part of a sensor network with a newly designed sampling probe to take samples from any point in wastewater tanks and basins. The new system supports a network of UAV platforms, whose design architecture is presented as a contribution, like its role as a sensor in a network sensor, the new custom sampling probe, and several results obtained during the development and testing of a prototype UAV. The rest of this work is structured as follows: the first section describes with more detail the sampling problem, the reference use case scenario considered for this work, and describes from with a high level point of view how the system works. The following section, "System and Hardware Structure" describes which components build the systems, how they are organized in subsystems, and put emphasis in the built UAV prototype, detailing its hardware and the organization in terms of connection and interfaces. Next section, "System design and software architecture" deals with the design of the system from the MAS point of view, how the software is (from a logical view) organized and implemented, and provides justification when the logical designs does not fit the actual implementation. This is followed by the "Experimental results" section, where the results obtained during the development and testing phase are discussed, especially for the localization problem of the unmanned aerial vehicle; in this section the resulting sample collector probe is presented and discussed as well. To conclude, the work presented is discussed in terms of the contributions made, and the most promising lines of work are commented.

2 UAV Network Sampling of Wastewater Plant

2.1 Wastewater Plant Sampling

In terms of management, one of the most critical processes to properly monitor a wastewater treatment plant is the evaluation of processed water, to ensure its compliance with the required specifications. The factors to be accounted and monitored include from chemical properties of the water, like pH, to concentrations of biological and inorganic pollutants. This frequently means obtaining multiple samples, with variable specifications (e.g.: samples for microbiological testing cannot be exposed to sunlight, nitric nitrogen samples must be processed within 2 h, etc....) [11–13]. Periodic testing requirements in this case means planning account for both the requisites and the resources available, be it technical or human.

For most of the metrics to test, both for clean waters and wastewaters, one of the most desired features is the homogeneity of the water to be sampled. This guarantees that the results obtained are representative of the water present in a given basin or tank, providing validity to the analyses. Notice that process tanks in water treatment plants present slow dynamics, with low flow velocities that drag dilution processes along all the tank. While these processes help homogenize the concentration of several pollutants and equalizes chemical indicators (increasing homogeneity of the sample and accuracy of the test), the low speed makes the tanks and basins to act as buffers. This fact produces delays hard to model, where water subjected to different rates of treatment and dilutive processes due entering the tanks at separated times can be present concurrently in the tank. In turn, this variability in the state of the water means that proper representative sampling of a single basin may require several samples at different points. This requirement may complicate the sampling process, as the optimal points for sampling from an analytic point of view may be hard to access, requiring special equipment or structures to reach them.

These features of the sampling process may present additional challenges for human workers beyond the accessibility of the areas, as in certain environments it is possible to find noxious gasses and substances, the samples may present special requirements (as commented above, with respect to timing, exposition, insulation...), and the weather or other factors can introduce further hindrances (excessive temperatures, insufficient lighting, dangerous conditions due wind speeds...). All these factors increase the complexity of the sampling procedures and the risks associated, especially for human operators. These complexities and risks may be managed with improvements in methodologies, resources and installations, but at the end they may only reduce risks, not remove them as long as the sample collection is performed by human workers.

2.2 Network Sampling Environment

For the proposed network sampling system, a medium sized wastewater processing plant is considered (see Fig. 1). The plant of Sant Feliu (Barcelona) services 320,000 inhabitants and relevant infrastructure and economic activities. It can process up to 72,000 m^3 of wastewater, coming mainly from 2 sources: domestic sewage, originated at residential housing and commercial areas, without special pollutants; and industrial wastewater, pre-treated by the industries to remove specific pollutants, and returned afterwards to the conventional wastewater circuit.

Fig. 1. Delayed Inverse-Depth (DI-D) Monocular EKF-SLAM Satellite view of the wastewater plant considered, located 10 km south-west from Barcelona, in Sant Feliu municipality. The plant compound measures 480 m long by 192 m wide (enclosed in light blue line). The open air basins area, where the network sampling system is designed to operate, is enclosed by the dashed green line, making 250 m long by 118 m wide. The basins present several sizes, with lengths varying between 35 m and 65 m, and widths in the range between 11.5 m and 13.5 m.

Water processed by the plant is expected to reach the quality metrics required by regulatory bodies to allow its return to the environment. Several processes are applied successively, from mechanical filtering and decanting, to acceleration of organic matter degradation. As part of the resources required for these processes, 17 outdoor open air tanks and basins are present, with an accumulated surface over 11,200 m^2.

2.3 UAV Network Sampling

The proposed system offers a solution for the sampling and the analyses required to monitor and supervise the open air basins in a wastewater treatment plant. The sampling network allows to analyse the water present in almost every point of the open basins and tanks, thus, acting as a virtual network of multiparametric sensors with hundreds of probes deployed. To achieve this goal, the robotic UAV platforms are fitted with the necessary probing equipment: either a sample collector probe which picks water samples around ~ 400 cm^3 to bring them to a delivery station; or a multiparametric probe sensor to sample the available parameters online while in flight.

Thus, a human controller who operates the system is able to set a list of sampling tasks. Each one of these tasks is composed of the coordinates of a point in one of the basins, where the sampling has to take place, and a set of instructions or restrictions related to the sampling, including which parameters are desired to measure, what time or period should the measurements or samples to be taken, whether is required to keep the sample or a single measurement is enough, and any other features. The sampling system then checks the available UAV sampling platforms, receiving from them updates on their status (including values as battery charge available), and tries to plan a sampling solution which satisfies the list of sampling tasks. For a mission where there is no need to do laboratory work, i.e., the metrics to be analysed can be obtained with the multiparametric sensor probe and there is no requirement to keep or validate the sample at a later point, a UAV platform with the multiparametric probe will be sent, and the results from the analysis will be provided through a VPN over a 4G link even before the UAV has returned and landed (see Fig. 2, left).

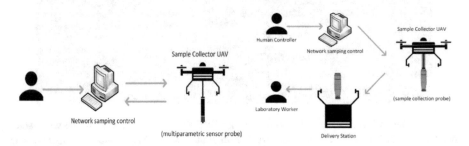

Fig. 2. (Left) UAV network sampling system for sampling missions using multiparametric probe. (Right) UAV network sampling system for sample capture mission

When there are conditions that makes the use of a multiparametric probe unsuitable (measurements that require long time to be measured, for instance hours or days for the consumed oxygen, or when the normative requires a contrast analysis with the same sample, etc....), a similar UAV with a sample collector probe (described in Water Sampling with a UAV section) will be sent. The operations of these UAVs are more complex, as the deployment of a probe that can be empty or filled with water, modifying the dynamic behaviour of the UAV. To fill it, the UAV submerges the probe into the open air basin by floating above the water, as the sample collection probe is rigidly solidary with the UAV. After enough time to fill the probe has passed, the UAV flies to a delivery station, where it releases the magnetic valve which sustains the probe, and also can land to be serviced (Fig. 2, right).

3   System and Hardware Structure

The system proposed is largely based in off-the-shelf (OTS) commercially available technologies for hardware, and open source software, reducing development and operational costs for the deployment of the prototype. This system is managed from a central control system, which is deployed in an ordinary PC, thus allowing to manage a network of sample collector UAVs and localization beacons. A single sample collector UAV prototype has been developed and built, based on a custom-design of UAV, with high power-to-weight ratio in order to be able to operate moderately heavy payloads affected by dynamics forces.

The sample collector UAV was designed as a quadcopter12 at X4 configuration, due its stability and performance, with a length of 0.664 m between rotor axes. The propellers are standard 18" build in carbon fiber, with T-Motor MN4014 actuators. These rotors are controlled at low level by a set of 4 T-Motor AIR 40A electronic speed controllers (ESC), which are connected to a standard power supply board paired to a Pixhawk 2.4.8 flight management unit (FMU). The power supply board is connected to a custom power system, feed by two 6S 10,000 mAh batteries. The custom power system allows several functions, like battery switch, balanced battery charge, and producing different voltages with a set of buck converter circuits.

The Pixhawk FMU is connected to a EGNOS SBAS13 enabled Ublox GPS, an internal IMU, a Lightware SFC10/B LRF set as altimeter, LRF two radio transceivers: a 915 MHz radio telemetry module; and a Futaba 2008RB 2.4 Ghz to enable direct manual control during experimental validation of newly developed features. To run all the sensors and hardware the Pixhawk is set with the PX4 flight stack14, providing all the basic flight capabilities, included some autonomous flight capabilities, like a return-to-home (RTH) function, and take-off and landing using a Tarot landing set. This quadcopter setup has a maximum take-off weight of 13.9 kg, with a payload weight ranging between 3.9 and 4.55 kg depending on the presence and configuration of a safety cage.

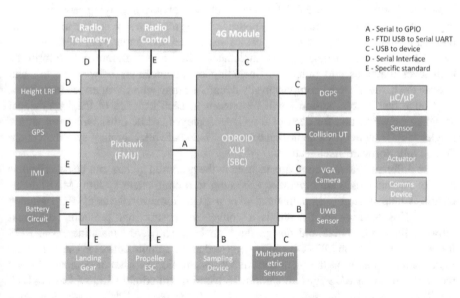

Fig. 3. Hardware architecture of the prototype sample collector UAV developed. Not all the described components will be used concurrently or during standard operations.

This base UAV was enhanced with robotic capabilities after the addition of several sensors and an Odroid X4 single board computer (SBC). This SBC deploys a Exynos 5522 octacore based on ARM designs (comparable to a not so new high end smartphone in computational power), and presents several USB and serializable interfaces, including a GPIO pinout. This allows to connect several sensors and devices, including: a USB 3.0 camera (up to 5MP, usually set a VGA resolution to ease bandwidth usage during monitoring), a SBAS (EGNOS network) receiver, a network beacon transceiver, and a set of 4 MaxBotix ultrasonic sensors to detect near-collisions and obstacles. All this additional hardware allows fully autonomous flight operations in known prepared environments. The final addition to the UAV system is the sampling system, with two different options: deployment of a multiparametric probe or the deployment of a custom sample collector probe, described at a later section. A full schema of the computing architecture and all the devices connected can be seen in Fig. 3.

Given the complexity of the system described, it is useful to consider the distributed sampling network developed in terms of a set of different subsystems, with each subsystem comprised of both the hardware and software components responsible for a task. The hardware level components are generally organized according physical restrictions and specifications, while at software level, a multi-agent system (MAS) approach, based on the methodology described in [14], was considered the most convenient approach. Using a MAS design abstraction allows produce software level components with great modularity and robustness, which are integrated with the hardware level components through a set of low-level routines and drivers, which build an interface layer. In terms of software implementation, the multi-agent architecture uses a framework based on ROS middleware. Using a ROS framework allows quickly deploying new software, organized in executable units called "nodes", easing the issues of setting up and configuring communications between processes distributed in different machines. Thus, for each subsystem, one or more agents implemented in the ROS framework perform the high level functions and processes (including communications between them), and operate the hardware level components by communicating with the interface level layer.

Notice that several systems will be present in each instance of the hypothetical sampling network, i.e., for each of the UAV deployed; while other will work akin a singleton class, with only one instance for the whole network sampling system. System Design and Software Architecture.

The different software components used in the proposed system can be described as a set of Agents which are organized according to a multi-agent system (MAS) architecture, which is largely implemented over a ROS framework9 (using ROS Kinetic Kame). This Multi-agent system is a convenient abstraction, providing a virtual infrastructure to organize and design the high level software capabilities. The ROS framework is run in an ODROID SBC deployed at the sample collector UAV, and the PC machine/s present at the main control subsystem. For the sake of compatibility all the hardware using ROS runs an Ubuntu 16 based distribution of Linux, hosting both the ROS framework, the software agents and all the required drivers and low level routines for communication and actuation. The fight management unit also runs several pieces of software at agent level, those closely related with the avionics systems of the UAV, and usually closely tied to the low level control routines communicating with the hardware level. These two devices communicate through a serial port in the FMU connected to the GPIO pinout in the ODRIOD. This architecture allows for perception and high level processes and agents to use the resources of the more powerful SBC, while the agents and resources needed to keep the sample collector UAV flying safely are isolated in the FMU, thus ensuring their performance and availability.

Note that a ROS framework usually requires that a single "ROS master node" (being a node the basic execution unit that communicates with other nodes through ROS) acts as a "master agent" managing all the routing and initial steps of every communication, and providing a unified clock for the system. This centralized architecture presents a critical fail point in this feature, as failure of this "node" usually invalidates the whole network. Moreover, although this can be managed in certain network environments, it becomes an unsolvable issue when the communications between the different physical elements of the system cannot be guaranteed, i.e.: deploying the master "node" in the ground or user stations at the main control system would endanger any sample collector UAV unit in flight if the 4G communication becomes unviable; while deploying it in a sample collector unit can risk other units (if they were present) and render the main control subsystem powerless.

In order to avoid the commented issues related to a master "node" acting as master agent, several master agents will be deployed using the multimaster FKIE tool. This allows to deploy multiple ROS master "nodes", and managing synchronization and coordination between them. So, from the MAS point of view, there will be a central master agent and a local master agent per UAV. The main difference between these central and local master agents resides in that the central master agent ensures coordination between the main control subsystem devices and every available UAV, while the local master agents only coordinate with the central master agent, and not between each other. This architecture ensures that the local ROS framework of an UAV will remain operative even if the connection with the ground station is not available, while isolating each UAV's communications from each other, to avoid sharing unnecessary ROS coordination data between them (which can require a noticeable bandwidth consumption, as discussed in [16–18]).

4 Experimental Results

4.1 Water Sampling with UAV

The hardware design and software architecture described in previous sections consider two different available setup options to sample the open air basins: (1) using a commercial OTS multiparametric probe sensor; (2) deploying a custom designed sample collection device, able to hold over 400 cm^3.

The first option presents two main advantages: it is an existing technology, widely supported, already integrated into PC environments, and from a flight and control point of view, a hanging solid body with fixed mass is a problem already solved by research in with several strategies. The second option also presents its own weaknesses and strengths: deployment of a hardware system able to collect water samples from an UAV robotic platform is cheap under the right designs, and although it is not able to sample on the fly, it can be used in processes where the sample must be kept for periods of time, or requires any kind of processing before performing analytics tests. Note that the utilization of the sample collector probe implies that the mass of the UAV platform, and its distribution (so inertia) will vary dynamically during the collection and return flight, introducing disturbances in the low level control loops.

The proposed sample collection probe was developed and built in our laboratory. The probe is able to hold ∼0.4l according to the fill tests, with a buoyancy valve that will lock the containment part as it floats to its top when filled. The probe is attached to the UAV by a ferromagnetic plate at its top, which is locked into a magnetic gripper. This magnetic gripper works with negative logic for safety, so in case of failure, the probe will be locked and cannot be released. The probe has to be submerged for ¾ of its height, so that waters reach the opposing entry holes, whose position ensure that the flow velocities of entering water induce negligible torque. To empty the probe, a screwcap at the bottom is used, which also allow retrieving the buoyancy valve for cleaning and maintenance.

4.2   UAV Flight Validation System Identification

A critical step in building any UAV platform is the setting and configuration of the avionics controller parameter, which have to be tuned individually for each design. From a robotics point of view, accurate modelling of the dynamic behaviour of the base UAV platform is critical to develop a robotics platform, as it is required for accurate localization and navigation operations (including trajectory planning). In order to satisfy this requirements, several indoor flight test were performed, in a motion capture testbed using an Optitrack® system (see Fig. 4). This system allowed to accurately validate the performance of the system according to the avionics controller parameters, and capture enough data to produce an accurate enough dynamic model to use in the robotics agents (localization and navigation) through parameter identification.

Fig. 4. UAV prototype platform flying indoors in an Optitrack® testbed. The UAV is equipped with a carbon fiber cage with a light metal grid for safety reasons, as the system is still in development stages.

4.3 UAV Localization

For any autonomous UAV system one of the most complex issues is always the localization problem. Commercial solutions with high accuracy are available, but present several challenges due the trade-offs between weight, accuracy, performance and economical costs. Though differential GPS can produce the most accurate results, its use requires deployment of expensive equipment both in the UAV and as a part of infrastructure, and the rate of its data (5–10 Hz) is not enough to perform robust localization and navigation for UAVs. All the other GPS-based solutions, like SBAS augmentation with the EGNOS network available also share this specific issue. Notice that for the proposed system, especially for the scenario considering the utilization of the sample collector probe, a correct estimation of height is critical. As it can be observed after the experiments, neither of the GPS-based approaches produce an accurate enough height estimation to perform the required manoeuvres to sample the basins flying close to the surface. To solve this issue, an LRF (laser range finder) altimeter was deployed. The avionics implements a low level extended Kalman filter [17, 18] to integrate results of the EGNOS-enabled GPS with measurements from the internal IMU and the laser altimeter.

Though the results produced by the avionics system fusing readings from all the sensors, the availability of the EGNOS enhancement is not guaranteed, as it has been widely discussed in literature [19]. Moreover, while the avionics system can reduce the height error thanks to the altimeter, the accuracy in the planar positioning is not good enough for the landing and sample collector probe delivery tasks. To solve this, the additional localization tasks are performed at the localization agent level. Using data from the camera agent, a fiducial marker detection process is used to achieve centimetre level accuracy in marked areas, allowing proper landing and probe delivery operations. To deal with the eventuality of EGNOS shortage, a system of beacons based in the TI chip CC2530 has been tested [15]. This component implements a full SoC solution (system on a chip) for ZigBee-based applications, and its low cost and low-power consumption makes it ideal for developing inexpensive, easy to deploy and maintain beacon networks. This system would be deployed around the open air basin, to improve UAVs localization during EGNOS shortages. To perform the test, a set of 6 beacons was deployed in a planar configuration covering an area of 30 m per 15 m. These tests reported errors within a compatible range of those reported within previous literature [19–21] when accounting for differences in the scenarios. Still, active research in the field is producing new solutions [22] that could improve the results.

5 Conclusions

This works proposes a solution to the problem of automatizing the sampling and monitoring of a wastewater treatment plant. The solution proposed is the utilization of a network of sampling UAVs with different sampling methods and capabilities. This UAV sampling network would act as a virtual sensor network composed of hundreds of multiparametric sensors/sample capturing probes, as it is capable of operating at any point of an accumulated area over 11,200 m^2. The sensing UAV network system proposed is designed as a multi-agent system, built over a ROS framework, which provides several advantages derived from the fact that each agent is responsible of a task and has its own status and agency, producing increased robustness, modularity, improved upgradability, and ease of maintenance.

With respect to the solution proposed as a system, several advantageous contributions over author's previous works have been presented. First of all, the proposed system is much better from a design and implementation point of view, with better isolation between tasks. The system architecture is fully scalable, so new sampling UAV can be added to the sampling network system with ease, as the MAS design allows to accommodate them with minimal changes in the implementation. Replicating the UAV sample collector is a trivial task, and migrating the MAS system into an upgraded hardware platform should not be complex, with the only challenges present in the mechanical integration of sampling devices.

In terms of the sampling process, the designed probe has been tested in simulated tanks, and its performance meets the specifications. The design with two entry points allows to reduce possible torques to negligible values, and allows the correct filling of the sample container. The buoyancy valve presents no losses under normal operation, and under shock circumstances, losing a sample is irrelevant, given other risks incurred during an UAV crash.

The actual performance of the robotic system has been only tested in terms of localization. Being one of the hardest challenges for UAV, achieving a satisfactory solution in terms of accuracy and performance without expensive equipments (be it in economical or payload terms), is always complex. Several solutions have been tested, with validation data provided by a highly accurate DGPS processed off-line. Tests revealed that SBAS-enhanced GPS with the EGNOS network can provide accurate enough data, but is not reliable enough to automate a system, so an alternative beacon method was tested. The beacons present slightly less accuracy, but results are much more robust, and are not subject to unpredictable disturbances. The biggest issue that remains to be addressed in the localization problem is replacing the laser range finder LRF altimeter by a device able to operate properly over the water, and reduce its impact.

Future works must address some specific challenges still pending, starting with a robust flight controller able to maintain the UAV over water while the sample collector fills the bottle with water to be analysed. The risks implied by using a multiparametric sensor probe can be easily managed at the hardware integration step, as the sensor can hang as much as needed, acting as an inverted pendulum (which has been solved for years). This specific challenge presents a strong dependency with the localization problem, as the controller must account for which specifications are guaranteed by the localization agent in terms of accuracy, rate and robustness. Finally, the sampling scheduler part of the sampling network manager agent can be further developed, as of now its implementation assumes that the UAVs are uniform except for the specific sampling device, and its performance is also uniform (e.g., it could not account that batteries and other components degrade with time dealing different performances and capabilities).

Acknowledgments. This research has been funded by AEROARMS EU Project H2020-ICT-2014-1-644271.

References

1. Heyer, C.: Human-robot interaction and future industrial robotics applications. In: 2010 IEEE/RSJ International Conference on Intelligent Robots and Systems (IROS), pp. 4749–4754 (2010)
2. Ping, J.T.K., Ling, A.E., Quan, T.J., et al.: Generic unmanned aerial vehicle (UAV) for civilian application-a feasibility assessment and market survey on civilian application for aerial imaging. In: 2012 IEEE Conference on Sustainable Utilization and Development in Engineering and Technology (STUDENT), pp. 289–294 (2012)
3. Loh, R., Bian, Y., Roe, T.: UAVs in civil airspace: safety requirements. IEEE Aerosp. Electron. Syst. Mag. **24**, 5–17 (2009)
4. Freeman, P., Balas, G.J.: Actuation failure modes and effects analysis for a small UAV. In: 2014 American Control Conference, pp. 1292–1297 (2014)
5. Niazi, M., Hussain, A.: Agent-based computing from multi-agent systems to agent-based models: a visual survey. Scientometrics **89**, 479 (2011)
6. Iñigo-Blasco, P., Diaz-del-Rio, F., Romero-Ternero, M.C., et al.: Robotics software frameworks for multi-agent robotic systems development. Robot. Auton. Syst. **60**, 803–821 (2012)
7. Conte, G., Scaradozzi, D., Mannocchi, D., et al.: Development and experimental tests of a ROS multi-agent structure for autonomous surface vehicles. J. Intell. Rob. Syst. **92**, 705–718 (2018)
8. Langley, P., Laird, J.E., Rogers, S.: Cognitive architectures: research issues and challenges. Cogn. Syst. Res. **10**, 141–160 (2009)
9. Quigley, M., Conley, K., Gerkey, B., et al.: ROS: an open-source robot operating system. In: ICRA Workshop on Open Source Software, p. 5 (2009)
10. Guerra, E., Bolea, Y., Grau, A., et al.: A solution for robotized sampling in wastewater plants. In: IECON 2016 - 42nd Annual Conference of the IEEE Industrial Electronics Society, pp. 6853–6858 (2016)
11. Urban Waste Water Directive - Environment - European Commission (1991). http://ec.europa.eu/environment/water/water-urbanwaste/legislation/directive_en.htm. Accessed 28 Feb 2019
12. Mahony, R., Kumar, V., Corke, P.: Multirotor aerial vehicles: modeling, estimation, and control of quadrotor. IEEE Robot. Autom. Mag. **19**, 20–32 (2012)
13. Subirana, J.S., Zornoza, J.M.J., Hernández-Pajares, M.: GNSS Data Processing, 238
14. Meier, L., Honegger, D., Pollefeys, M.: PX4: A node-based multithreaded open source robotics framework for deeply embedded platforms. In: 2015 IEEE International Conference on Robotics and Automation (ICRA), pp. 6235–6240 (2015)
15. CC2530 Second Generation System-on-Chip Solution for 2.4 GHz IEEE 802.15.4/RF4CE/ZigBee| TI.com. http://www.ti.com/product/CC2530. Accessed 28 Feb 2019
16. Hernandez, S., Herrero, F.: Multi-master ROS systems. IRI-TR-15-1 (2015). http://hdl.handle.net/2117/80829
17. Zeng, J., Li, M., Cai, Y.: A tracking system supporting large-scale users based on GPS and G-sensor (2015). https://journals.sagepub.com/doi/abs/10.1155/2015/862184. Accessed 28 Feb 2019

18. Trujillo, J.-C., Munguia, R., Guerra, E., et al.: Cooperative monocular-based SLAM for multi-UAV systems in GPS-denied environments. Sensors **18**, 1351 (2018)
19. Liu, J., Chen, R., Chen, Y., et al.: Performance evaluation of EGNOS in challenging environments. J. Global Position. Syst. **11**, 145–155 (2012)
20. Adnan, T., Datta, S., MacLean, S.: Efficient and accurate sensor network localization. Pers. Ubiquit. Comput. **18**, 821–833 (2014)
21. Guo, Y., Liu, X.: A Research on the Localization Technology of Wireless Sensor Networks Employing TI's CC2530 Instrument. IEEE, pp. 446–449
22. Thammavong, L., Khongsomboon, K., Tiengthong, T., et al.: Zigbee wireless sensor network localization evaluation schemewith weighted centroid method. In: MATEC Web Conference vol. 192, p. 02070 (2018)

Identification of the Relationships Between a Vocal Attribute and a Personality

Kimberly Brotherton, Jangwoon Park$^{(\boxtimes)}$, Dugan Um,
and Mehrube Mehrubeoglu

Department of Engineering, Texas A&M University-Corpus Christi,
Corpus Christi, TX, USA
kimberly.brotherton@yahoo.com,
{jangwoon.park,dugan.um,ruby.mehrubeoglu}@tamucc.edu

Abstract. There is an increasing need for human interaction to improve with electronics in this digital age. Socially assistive robot (SAR) interactions with the elderly and youth can improve the quality of these individuals' lives in terms of friendships. Since humans are emotional creatures and respond more agreeably to similar personality types, robots need to be designed and programmed in a more intuitive way to capture and match the users' personal characteristics to maximize human-machine friendships. The present study is intended to identify significant vocal features associated with a human's personality type (introverted vs. extroverted), in a digital signal processing environment, and use vocal traits for characterization. The voices of 28 university students (14 introverted and 14 extroverted) were recorded when each verbally responded to the Walk-in-the-Woods questionnaire. Then, the response time for the first question for each participant was extracted. Statistical analyses were employed to test significances of each measure for the two personality groups (introverted vs. extroverted).

Keywords: Personality · Socially assistive robot · SAR · Vocal attribute

1 Introduction

Socially assistive robots (SARs), such as Jibo [1], Pepper [2], Baxter [3], and Hospi [4], are currently being developed to improve people's quality of life. According to the Bloomberg Businessweek, Robosoft Technologies plans to produce 10,000 Kompaï robots annually to assist seniors at home by 2020 [5]. The SAR interacting with the elderly, for example, can improve the quality of life of these persons in terms of cognitive care, rehabilitation, and companionship. Since humans are emotional creatures and respond more agreeably to similar personality types [6], robots need to be designed and programmed more intuitively to capture and respond to the users' personal characteristics to maximize human-robot friendships.

Several studies have been conducted to promote the human-robot interaction by changing a SAR's behavior toward a user's personality. Andrist *et al.* (2005) analyzed users' gaze behaviors to identify their individual personality so that a SAR can behave differently toward the identified personality [6]. As a result, the SAR promoted engagement and provided motivation to the users to conduct a therapeutic task.

© Springer Nature Switzerland AG 2020
J. Chen (Ed.): AHFE 2019, AISC 962, pp. 163–167, 2020.
https://doi.org/10.1007/978-3-030-20467-9_15

Goetz *et al.* (2003) found the positive effect in eliciting user compliance by changing the robot's behavior toward the users [7]; therefore, we can conclude that, by matching a robot's behavior with an individual human, a social robot can respond with a more appropriate interaction.

Several studies have identified objective measures to estimate a user's personality such as facial images from Facebook profiles [8–11]; however, research in classifying the two personalities (introverted vs. extroverted) based on speech data, such as how quickly they response a question or how long they answer, is limited.

The present study is intended to identify significant vocal features associated with a human's personality type (introverted vs. extroverted). The novel personality classification models will be developed by incorporating significant vocal features with advanced modeling techniques.

2   Methods

In this section, selection of participants, devices and procedure for collecting voice data, and data analysis are presented.

2.1   Participants

Twenty-eight native-English speakers with different personalities participated in this study. To recruit similar numbers of introverted and extroverted participants, a pre-screening process was conducted by checking a potential participant's Myers-Briggs Type Indicator (MBTI) personality via an online tool. The MBTI is one of the most popularly used in the academic research area on personality [12]. The test is taken in questionnaire form and identifier letters are assigned to the 16 personality types, where eight types belong to introvert and the other eight types belong to extrovert. For example, if a participant was classified as INFJ out of the 16 personality types ("I" stands for introversion, "N" stands for intuition, "F" for feeling, and "J" for judging), then he/she was classified as an introverted person; if a participant was classified as ENFJ ("E" stands for extraversion), then the person was classified as an extroverted person. In this study, we recruited 14 introverted (1 female and 13 males) and 14 extroverted (2 females and 12 males) participants based on their MBTI personality types. Most of the participants were university students at the age of 20s.

2.2   Apparatus

To measure the participants' voices, a Walk-in-the-Wood questionnaire and an audio recorder (Samsung Electronics Co., South Korea) were used. The Walk-in-the-Woods relational psychology test is a fun test that was administered by one interviewer at the same talking speed to each subject. This test consists of eight questions including "Picture yourself walking through a beautiful forest. The sun is out, there's a perfect breeze. It's just beautiful. Who are you walking with?" and "As you continue in your walk through the forest, you come across an animal. What kind of animal is it?" These questions indicate relevance to values that the data pool subjects deem important in their personal lives [13].

2.3   Experimental Procedure

The data collection was conducted on the university campus, at a laboratory setup, at a time that works for the participants and moderator. At the beginning of the experiment, the purpose of the experiment was clearly and thoroughly explained to the participants. Their informed consent forms were obtained. Next, the moderator used the Walk-in-the-Woods questionnaire to ask the participant's personal answer. When the moderator asked the Walk-in-the-Woods questions to the participant, then the participant made his/her answers right after the questions. Throughout the experiment, the moderator's and participants' voices were recorded. The experiment took up to 10 min to complete. The experimental procedure was approved by the Texas A&M University-Corpus Christi, Institutional Review Board (IRB, ID: 61–18).

2.4   Analysis Method

The captured sound was analyzed to quantitatively characterize the voice for identifying the relationship between sound and personality. The response times of the participants' answers were taken from the recorded audio files. The sampling rate of the recorded audio was 48,000 data points per second. To analyze response time of each participant, the voices of the moderator were manually removed from the recorded audio files. After that, each participant's response time were extracted in seconds. For example, Fig. 1 shows one introverted participant's voice data without the moderator's voice. The horizontal axis indicates recorded data points with a sampling rate of 48,000, and the vertical axis indicates amplitude of the recorded sound. In this study, the response time for the first question, or first response time (FRT) was measured. To conduct statistical testing, two-sample t-test method was employed to identify significant features with a significant level at 0.05.

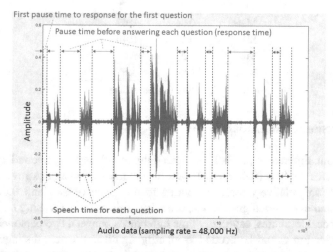

Fig. 1. Plot of a recorded audio of an introverted participant by using MATLAB. Note that the Walk-in-the-Wood questions consisted of eight questions.

3   Results

Table 1 shows FRT of each participant for the introverted and extroverted groups. The mean, standard deviation (SD), minimum, and maximum are 0.64 s, 0.44 s, 0.23 s, and 1.88 s, for the introverted group, and 0.62 s, 0.40 s, 0.21 s, and 1.67 s, for the extraverted group, respectively. As a result of statistical testing, we found that an average difference of FRT between two groups (mean = 0.64 s for the introverted group and 0.62 s for the extroverted group, $t(25) = 0.10$; p-value = 0.922) was not significant.

Table 1. Measured FRT for each participant in the introverted and extroverted groups. (unit: second)

Subject no.	FRT	
	Introverted	Extroverted
1	0.77	0.56
2	0.83	0.42
3	0.23	0.73
4	0.63	0.21
5	0.54	0.31
6	0.35	1.25
7	0.27	0.73
8	0.52	0.52
9	0.31	0.52
10	0.98	1.67
11	0.94	0.73
12	0.25	0.38
13	0.42	0.42
14	1.88	0.25
Mean	0.64	0.62
SD	0.44	0.40
Min	0.23	0.21
Max	1.88	1.67

Note: FRT = first response time.

4   Discussion and Conclusion

No significant difference of FRT between introverted and extroverted groups was identified, since the SDs (0.44 s and 0.40 s for the introverted and extroverted groups, respectively) are relatively large compared to the mean (0.64 s and 0.62 s for the introverted and extroverted groups, respectively). This study is focusing on identifying significant vocal attributes, especially time differences between two personality groups. More investigation would be needed to identify significant factors distinguishing two personalities, such as average response time, total response time, total speech time, pitch frequency for a vowel sound or sound pressure level.

References

1. Andrist, S., Tan, X.Z., Gleicher, M., Mutlu, B.: Conversational gaze aversion for humanlike robots. In: Proceedings of the 2014 ACM/IEEE International Conference on Human-Robot Interaction, pp. 25–32. ACM, March 2014
2. Bickmore, T.W., Caruso, L., Clough-Gorr, K.: Acceptance and usability of a relational agent interface by urban older adults. In: CHI'05 Extended Abstracts on Human Factors in Computing Systems, pp. 1212–1215. ACM, April 2005
3. Bruno, B., Mastrogiovanni, F., Sgorbissa, A.: Functional requirements and design issues for a socially assistive robot for elderly people with mild cognitive impairments. In: 2013 IEEE RO-MAN, pp. 768–773. IEEE, August 2013
4. Bull, R., Gibson-Robinson, E.: The influences of eye-gaze, style of dress, and locality on the amounts of money donated to a charity. Hum. Relat. **34**(10), 895–905 (1981)
5. Bloomberg Businessweek: Europe Bets on Robots to Help Care for Seniors (2016). https://www.bloomberg.com/news/articles/2016-03-17/europe-bets-on-robots-to-help-care-for-seniors
6. Andrist, S., Mutlu, B., Tapus, A.: Look like me: matching robot personality via gaze to increase motivation. In: Proceedings of the 33rd Annual ACM Conference on Human Factors in Computing Systems, pp. 3603–3612. ACM, April 2015
7. Goetz, J., Kiesler, S., Powers, A.: Matching robot appearance and behavior to tasks to improve human-robot cooperation. In: The 12th IEEE International Workshop on Robot and Human Interactive Communication. Proceedings. ROMAN 2003, pp. 55–60. IEEE, October 2003
8. Liu, L., Preotiuc-Pietro, D., Samani, Z.R., Moghaddam, M.E., Ungar, L.: Analyzing personality through social media profile picture choice. In: Tenth International AAAI Conference on Web and Social Media, March 2016
9. Al Moubayed, N., Vazquez-Alvarez, Y., McKay, A., Vinciarelli, A.: Face-based automatic personality perception. In: Proceedings of the 22nd ACM International Conference on Multimedia, pp. 1153–1156. ACM, November 2014
10. Back, M.D., Stopfer, J.M., Vazire, S., Gaddis, S., Schmukle, S.C., Egloff, B., Gosling, S.D.: Facebook profiles reflect actual personality, not self-idealization. Psychol. Sci. **21**(3), 372–374 (2010)
11. Celli, F., Bruni, E., Lepri, B.: Automatic personality and interaction style recognition from Facebook profile pictures. In: Proceedings of the 22nd ACM International Conference on Multimedia, pp. 1101–1104. ACM, November 2014
12. Furnham, A.: The big five versus the big four: the relationship between the Myers-Briggs Type Indicator (MBTI) and NEO-PI five factor model of personality. Personality Individ. Differ. **21**(2), 303–307 (1996)
13. Lawton, G.: What Is The Forest Personality Test? This Fun Trick Might Tell You Something About Yourself (2016). https://www.bustle.com/articles/167161-what-is-the-forest-personality-test-this-fun-trick-might-tell-you-something-about-yourself

Human, Artificial Intelligence, and Robot Teaming

Humans and Robots in Off-Normal Applications and Emergencies

Robin R. Murphy$^{(\boxtimes)}$

Department of Computer Science and Engineering, Texas A&M University,
TAMU 3112, College Station, TX 77843-3112, USA
robin.r.murphy@tamu.edu

Abstract. Unmanned systems are becoming increasingly engaged in disaster response. Human error in these applications can have severe consequences and emergency managers appear reluctant to adopt robots. This paper presents a taxonomy of normal and off-normal scenarios that, when combined with a model of impacts on cognitive and attentional resources, specify sources of human error in field robotics. In an emergency, a human is under time and consequences pressure, regardless of whether the mission is routine or whether the event requires a change in the robot, the mission, the robot's work envelope, the interaction of the humans engaged with the robot, or their work envelope. For example, at Hurricane Michael, unmanned aerial systems were used for standard visual survey missions with minor human errors but the same systems were used at the Kilauea volcanic eruption for novel missions with more notable human errors. An examination of two cases studies suggests the physiological and psychological effects of an emergency may be the primary source of human error.

Keywords: Human factors · Systems analysis · Human error analysis ·
Extreme environments · Uninhabited aerial vehicles

1 Introduction

Ground, aerial, and marine robots are being used at disasters, public safety incidents, and other non-routine, also called "off-normal" events. Unmanned systems for off-normal events are unlikely to be fully autonomous. As noted by Murphy and Burke [1], many missions are intended to provide humans with real-time remote presence, not taskable agency, and thus no advances in autonomy would eliminate the human. Even fully autonomous missions, such as photogrammetric mapping with unmanned aerial vehicles, still involve human supervision and a human is always on call in case of a problem. Although robot deployments have generally been successful, some failures do occur and human error is responsible for over 50% of those cases [2]. While not related to human error, we have witnessed firsthand a hesitancy by managers to deploy robots for off-normal events.

In order to reduce human error and foster appropriate adoption of robots for off-normal events, this paper poses a taxonomy of off-normal events as either novel or an emergency. The paper uses this framework to discuss a model of human resource

© Springer Nature Switzerland AG 2020
J. Chen (Ed.): AHFE 2019, AISC 962, pp. 171–180, 2020.
https://doi.org/10.1007/978-3-030-20467-9_16

constraints and demands on the human in off-normal events, inspired by multiple resource theory [3]. An off-normal event increases the resource demands due to changes in the robot, the mission, the robot and human work envelopes, or the humans' cognitive capacity, while at the same time decreases the available resources due to changes in the physiological and psychological state of the humans. The human factors between a novel event and an emergency is that an emergency always diminishes the capacity of the humans due to unfavorable changes in physiological and psychological state. The paper then presents two cases studies of the use of small UAS during emergencies, describing how the resources demands and state changed and the human errors. The paper concludes with a discussion of ramifications of the taxonomy and model for reducing human error and increasing trust.

2   Taxonomy of Normal and Off-Normal

There appears to be no formal definition of normal and off-normal use of robots in the literature, therefore, this paper provides the following definitions and taxonomy. A robot can be used in one of three types of events: normal, off-normal due to an emergency, or off-normal due to novelty.

An off-normal event due to an emergency is when a robot is being used to respond to or mitigate an abnormal situation that is declared by the responsible authority to be time-critical. Emergencies are generally extraordinary. A hurricane is an emergency, an emergency room in a hospital is not. In an emergency event, the robot may be used for a normative application, but the humans must perform under pressure from time and consequences. For example, a small unmanned ground robots (UGV) for bomb squads and military operations were used during the Fukushima Daiichi nuclear accident for missions routinely encountered in those applications (open a door, go inside, look around, pick and examine an object, etc.) [2], but a failure in completing the mission or taking too long to complete the mission would have significant consequences. More often than not, a robot is used during an emergency for a novel application. For example, during the Fukushima nuclear accident, a small unmanned aerial system (UAS) typically used to track vehicle movement in the desert was used for inspection of the reactor buildings [2]. Both examples illustrate how emergencies stress the humans interacting with the robot.

However, an off-normal event can be different due to novelty, that something has changed about the normal way of working. There are five ways in which a routine can be disrupted:

- *Robot:* The robot platform and its capabilities has changed. This may be due to the replacement of an existing robot with a new or upgraded robot, the addition of a payload, the change in software, etc.
- *Mission:* Either the objective or the set of tasks the robot will perform has changed. This could be due to a new mission or use for the robot. It can also be due to a new workflow for accomplishing an existing mission, such as adopting a new set of best practices. Since the applications for robots is still formative.

- *Robot Work Envelope:* The physical environment in which the robot will perform the mission has changed. In a building collapse, the physical environment is now spatially restricted and cluttered with debris.
- *Humans:* The composition of the operators, mission specialists, and any other humans directly involved in using the robot for the mission changes, either new people are introduced, or roles change, or the skills and cognitive capacity of the humans change. A robot operator may not be trained for a new mission or may be ill. A structural inspection mission may require an engineer who has never worked with the operators to help direct the robot.
- *Human Work Envelope:* The physical environment in which the humans directly involved in using the robot for the mission are working has changed. The changes might be due to weather (humans used to working in warm weather are now working in bitter cold), safety (humans have to wear gas masks or hazmat suits), or background distractions (humans have to work in an unusually noisy or distracting setting).

The first three factors on the list above— Robot, Mission, and Robot Work Envelope– are generally referred to as the ecology of the robot in behavior-based robotics [4]. The other two factors capture how the human fit within the larger human-robot ecology.

3   A Model of Cognitive and Attention Resources in Field Robotics

The taxonomy presented in the previous section identifies types of influences on the overall human-robot interaction in a robot but falls short in connecting those influences to human error. This section introduces a model of cognitive and attention resources for field robots, describes how the five attributes of a human-robot system contribute, in conjunction with physiological and psychological drains, to cognitive and attentional deficits, which in turn increase the potential for human error.

Our model groups the drain on the cognitive and attentional resources for a human directing a robot into three broad categories: task demands, perceptual demands, and team work demands. These are described as follows.

Task demands stem from how the robot is used to accomplish the tasks. A human can be presented with a new robot, upgrades or software that changes how a task is performed. A new mission can either uses existing tasks and skills but in different ways or it can require new tasks and new skills. A mission can be performed under notably different environmental conditions, such as flying a small UAS during the day and flying the same mission at night. Using the taxonomy, changes in task demands would be seen as changes in the Robot (change in platform, sensors, or software), Mission (changes in tasks), Robot Work Envelope (changes in environment), and Human (changes in requisite skills).

Perceptual demands stem from the challenges of perceiving a distal environment, and actions upon said environment, that is mediated by a robot and interface. The human users are trying to comprehend an environment that may have changed and is being viewed from unfamiliar angles, such as from very near to the ground or a bird's eye view. Perception of the distal environment is further impacted by the design of the user interface, which may not display all relevant data or display it poorly. Using the taxonomy, changes in perceptual demands would be seen as changes in the Robot (change in sensors or sensor placement), Robot Work Envelope (operating in a deconstructed environment), and Human Work Envelope (change in the user interface).

Team work demands stem from the interaction between the human and robot, but also the interactions between multiple humans. Many robot missions benefit from multiple humans working together, for example, experiments show that two responders using a single robot to find a victim are nine times better than a single responder using the robot [5]. Regulations may require multiple operators, such as Federal Aviation Regulations for visual observers to assist pilots in safe UAS operations. An off-normal event is often characterized by additional personnel, for example, experts or decision makers looking over the shoulder of operators, as seen at Hurricane Harvey [6], or operators working in teams rather than solo during training to maximize teaching efficiency. Using the taxonomy, changes in team work demands would be seen as changes in mission and in humans.

Following Wickens Multiple Resource Theory [3], if the sum of these demands exceed the current capacity of the human, then increases in human error are expected to result. However, off-normal events may decrease the inherent capacity of the human through physiological and psychological effects. In an emergency, responders, including robot operators, get very little sleep, perhaps only 3 h power naps [7]. The operators may be asked to work a 24-h or longer shift, as we witnessed at the 2018 Kilauea volcanic eruption where a small UAS crew worked a full day shift but were then called out for a night shift. The responders have less quality of rest, essentially camping out in harsh conditions. In addition, the graveness of emergencies imparts a psychological burden on responders; it is impossible not to think about the disaster and its impacts or not to worry about the robot's performance [7].

Figure 1 provides a graphical metaphor. The outer concentric circle represents the absolute fixed amount of an individual human's cognitive and attentional resources. The inner circle represents the actual resources consumed to meet the task, perception, and team work demands. If the task, perceptual, and team work demands (arrows radiating out) are low, shown in green, then human error due to these demands (as opposed to human error for other reasons) would be less likely. If the demands are high and approach the human's limits, shown in red, the cognitive pre-conditions for human error will increase. However, the size of the outer circle can be diminished due to physiological and psychological pressures (arrows directed inward). This means that even if the task, perceptual, and team work demands remain the same as for a normal event, it is possible that physiological and psychological pressures of an off-normal event can drive the human into a region of high likelihood of making an error.

The model of cognitive and attention resource constraints and demands helps solidify the differences in potential for human error between novel and emergency events. In both an emergency and novel event, one or more of attributes of a robot system (the robot, mission, robot work envelope, humans, or human work envelope) changes. These changes may or may not lead to an increase demand on cognition and attention that exceed resources. A novel event may have some physiological and

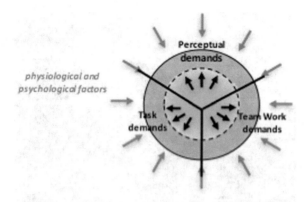

Fig. 1. A model of cognitive and attention resource demands in field robotics.

psychological costs that decrease the available resources, but in general, since it is not an emergency, the humans would be expected to be well-rested, in good health, can arrange to be in a work-conducive setting, and can control when they take breaks and pace of work. However, in an emergency, the humans operating the robot will always experience physiological and psychological deficits due to the nature of emergencies. These deficits go beyond apprehension of trying something new or being inserted into a new group of people or location.

4   Two Case Studies of Emergencies

Our participation as part of the Center for Robot-Assisted Search and Rescue (CRASAR) for its two most recent deployments to federally declared disasters provides case studies that illustrate how emergencies increase the likelihood of human error. CRASAR has participated in 30 disasters since 2001, for the sake of space, only the two are discussed.

4.1   2018 Kilauea Volcanic Eruption

CRASAR deployed to the Kilauea volcanic eruption to assist the Hilo, Hawaii, Department of Civil Defense with small UAS from May 14 to May 19, 2018. The team flew a total of 28-day time flights and 16 night time flights to fulfill three types of missions:debris/damage/flood estimation, strategic situation awareness/reconnaissance/survey, and tactical situation awareness. The team did not have a predictable schedule and worked more than 12 h on May 14 and more than 24 on May 17–18. The team had to carry, and often wear, respirators to protect from SO2 emissions as shown in Fig. 2. The human work envelopes were generally stationed close to the volcano, on the order of 0.1 mile to 0.25 miles; this was considered outside of the range of lava and debris explosions, but the noise was deafening.

CRASAR provided four models of small UAS and five pilots, three of whom had flown at multiple disasters and all but one was cross-trained on all models. The team

Fig. 2. Pilot and visual observer wearing respirators during exposure to SO2 at Kilauea volcanic eruption.

members were all proficient with the robots and had flown together in exercises. The team divided into three squads, with two of the squads consisting of a pilot, visual observer, and data manager, and one a pilot and visual observer. The team members had flown from Texas, Florida, and Utah, and arrived with jet lag from long flights and the three- to five-hour time difference. The squads travelled approximately 1.5 h between the affected areas and a rented house, where team members each had a bed and most had a private room.

The high potential for human error was observed on one occasion. On the night of May 14, the team was given a routine survey mission, requiring the use of a new DJI Zenmuse XT2 thermal sensor on the DJI Inspire. The pilot had not used the Inspire in over a month and had to fly the first mission at night from a small landing zone surrounded by difficult to see power lines. The Inspire evinced a platform problem on the first flight, was landed successfully and fixed. On the second flight, the platform encountered electromagnetic interference to the compass and inertial measurement unit which caused the UAS to randomly change direction. This forced the pilot to put more effort into navigation. The mission was successfully completed. Note that Robot had changed from normal operations (new sensor), Robot Work Envelope had changed (more constrained than normal, night), the Humans had changed (skills were rusty), and the Human Work Envelope had changed (wearing gas masks, night, noisy explosions).

On the night of May 16, the team was given a routine mapping mission from a previous site. It was flown with the DJI Inspire but now operations were routinized. The mission had to be repeated three times as the platform did not correctly record imagery on the first two flights. The cause was presumed to be human error: that the data manager or pilot had done something wrong in setting up the autonomous software or the data transfer process and, through increasing the level of detail of the methodical check and confirmation of each step in the process, finally performed the correct sequence. While this could have been a software error, the error never reappeared.

Note that the only change from the previous two days was the humans' degraded physiological condition from lack of sleep and rest.

During the day of May 17, one of the pilots had a problem with a software program that allowed autonomous image collection; it would not reach altitude and collect pictures but would simply hover. Eventually the pilot determined that the altitude limit from a previous flight in the continental US had not been reset on the configuration menu to the appropriate limit for the volcanic event. The autonomous program then performed correctly. Again, the only change was the humans' degraded physiological condition from lack of sleep and rest.

4.2   2018 Hurricane Michael

Hurricane Michael made landfall on October 10, 2018. CRASAR, under the direction of the Center for Disaster Risk Policy, a member of the State of Florida Emergency Response Team, deployed to Hurricane Michael with small UAS. The team flew a total of 80 flights to fulfill 26 missions from Oct. 11 to Oct 14. at Panama City and Mexico Beach. Small UAS were used for three types of mission: debris/damage/flood estimation, strategic situation awareness/reconnaissance/survey, and ground search. The squads flew a predictable schedule, from early morning to sun down each day, and conducted one mission at night, and from much less restrictive work envelopes (see Fig. 3).

Fig. 3. Typical human work envelope at Hurricane Michael.

The deployment was different Kilauea in many ways, most notably that the response had a predictable pace. The model of robots used were different, but the pilots were expert in their use. The missions, with the exception of ground search for missing people, were identical to Kilauea, but more of the missions involved photogrammetric mapping surveys. The robot work envelope was smaller, as temporary flight restrictions limited flights to 200 feet AGL versus 1,000 feet AGL. There were 11 pilots forming

6 squads, everyone on a squad knew each other and had flown together. Four of the five pilots from Kilauea participated, along with two other pilots who flew regularly with CRASAR. However, five additional pilots had not flown with the CRASAR pilots before, and only one of the four had post-disaster flying experience. Two of the eight pilots were unfamiliar or rusty with the DroneDeploy photogrammetric mapping software for survey missions. Four of the pilots traveled from Texas, 2 pre-deployed and the other two drove for about 18 h after the landfall. Eight of the team slept on cots in the Walton County Emergency Operations Center, about a two-hour drive from the affected areas; the rest stayed in hotels similarly far away. The teams were divided into 6 squads. Five of the squads included a Pilot and usually a Visual Observer due to high density of manned aircraft and other UAS from news organizations and from unauthorized flyers. If a squad had multiple pilots, they turns flying to give each other rest breaks.

Human error was observed on two occasions. Several squads had task errors where they mislabeled data, which was corrected by the data manager. The error was possibly due to lack of familiarity with the protocols, but the need to finish the flights before night fall and the distracting excitement of participating in a disaster, plus fatigue, probably heavily contributed to disregarding the protocols.

A more serious error was the generation of an overlap in the areas of two different UAS and the failure of one squad to detect a potential midair collision. In this situation, one of the two squads working in adjoining areas specified a region for their UAS to map that overlapped the region assigned to the other squad. By coincidence, both squads specified the starting point for the flight paths such that both UAS happened to be flying in the area of overlap at the same time at the same altitude. Fortunately, one squad saw the other UAS and changed altitude and aborted their mission to avoid a collision. It was unclear that whether the other squad was aware of the proximity of their UAS to the other. The squad that detected the problem was more experienced that the team in the adjacent sector. Note that the Robot Work Envelope had changed from normal situations (other UAS in the vicinity, which is uncommon) and the Humans had changed (one squad was not experienced).

5   Ramifications and Recommendations

Off-normal operation of robots should be of high interest to the human factors community because these deployments have high consequence if they are for an emergency and because off-normal conditions occur frequently due to upgrades, new features and best practices, and the identification of new applications for unmanned systems.

In terms of minimizing human error in emergency off-normal events, there are at least three non-exclusive approaches. One is to pro-actively minimize the novelty of the human-robot system being deployed. Three possible methods are below:

- Humans should have expert skill in normal conditions and, ideally, experience or training in novel and emergency conditions.

- Deployments should avoid the insertion of new technology, software updates, or anything else that could increase the potential for human error. This may not be possible. For example, the use of the XT2 and the toxic gas sensors were essential to the missions. However, these were minor modifications. One of the pilots had experience with the XT2 so it was not particularly novel and the toxic gas sensor was flown in benign conditions. On the other hand, during Hurricane Harvey, a UAS manufacturer put out a mandatory software update which introduced a dangerous bug.
- A "warm-up" area should be set aside to allow operators to refresh rusty skills or try out a new sensor before the actual mission. This may not be possible due to the location of the event or time pressures.

A second approach is to minimize unfavorable conditions for the humans. Perhaps the most obvious is to specify crew rest requirements. However, any such policies may have to be violated due to the exigency of the emergency. Sometimes responders simply have to work day and night. Also, the risk of a robot failure may be acceptable; the loss of a platform over a lava field is a relatively small monetary loss versus the reward of safely evacuating civilians.

A third approach is to mitigate the unfavorable conditions for the humans. Mitigations might be:

- Training individuals and teams for resilience.
- Minimizing the number of new members in a squad so as to preserve established norms.
- Require crew resource management protocols, such as having checklists, mission rehearsal, verbal protocols, a "sterile cockpit" with no unnecessary conversation, keeping observers or non-essential personnel out of the personal zone of the pilot to prevent distractions, and so forth.
- Increased autonomous capabilities, but these should be normative rather than only for use when the human's cognitive and attentional resources are depleted. The human may not trust a transfer of the autonomy and supervising the autonomous capability may further overtax the team. Also, if the autonomy was sufficiently good to be used to replace a person, why would it be used only in extraordinary situations?

A related topic is fostering trust in the unmanned system by explicitly considering whether the off-normal event significantly changes the normal use patterns of the robot. To date, managers are reluctant to use a robot for an off-normal situation. However, the use of a robot for a situation where it is performing an identical mission and a similar work envelope with trained, rested operators working in favorable conditions would not be expected to increase the frequency or severity of human error. The likelihood of a mission failure would be similar to the likelihood of a failure for the normal case. Given that the robot is used for normal events, the risks should be well understood.

The growing use of unmanned systems, and the continuous beta product development cycle, means that off-normal events will become more prevalent. This is an opportunity for the human factors community to investigate how to design autonomous capabilities or decision support for off-normal events, what training will produce resilient operators, and what mitigations are effective (crew rest schedules, checklists, crew resource management, protocols, and so forth).

Our current research efforts, with colleagues Drs. Ranjana Mehta and Camille Peres at Texas A&M, are focusing on quantifying the difference between normal and off-normal events, analyzing data from CRASAR deployments to inform crew rest requirements and mitigations, and creating resilience training for robot operators.

Acknowledgments. Portions of the work discussed in this report were funded by grants from the National Science Foundation (CNS 176047) and the Department of Energy (DE-EM0004483). We thank Odair Fernandes for his help in the preparation of this manuscript.

References

1. Murphy, R.R., Burke, J.L.: From remote tool to shared roles. IEEE Robot. Autom. Mag. **15**(4), 39–49 (2008)
2. Murphy, R.R.: Disaster Robotics. MIT Press, Cambridge, MA (2014)
3. Wickens, C.D.: Multiple resources and mental workload. Hum. Factors **50**(3), 449–454 (2008)
4. Arkin, R.: Behavior-Based Robotics. MIT Press, Cambridge, MA (1998)
5. Burke, J.L., Murphy, R.R.: Human-robot interaction in USAR technical search: two heads are better than one. In: 13th IEEE International Workshop on Robot and Human Interactive Communication, pp. 307–312 (2004)
6. Fernandes, O., Murphy, R., Adams, J., Merrick, D.: Quantitative data analysis: CRASAR small unmanned aerial systems at Hurricane Harvey. In: IEEE International Symposium on Safety, Security, and Rescue Robotics, pp. 1–5 (2018)
7. Casper, J., Murphy, R.R.: Human-robot interaction during the robot-assisted urban search and rescue response at the World Trade Center. IEEE Trans. Syst. Man Cybern. B Cybern. Part B **33**(3), 367–385 (2003)

Conveying Robot State and Intent Nonverbally in Military-Relevant Situations: An Exploratory Survey

Lauren Boos[1] and Lilia Moshkina[2(✉)]

[1] United States Military Academy, West Point, USA
Lauren.Boos@westpoint.edu
[2] SoarTech, Ann Arbor, USA
Lilia.Moshkina@soartech.com

Abstract. To successfully integrate into society, today's robots, largely non-anthropomorphic, need to be able to communicate with people in various roles. In this paper, we focus on the military robot domain, where radio silence and high cognitive loads are par for the course, making nonverbal means of communication a necessity. To better understand how the findings in nonverbal communication using visual (lights) and auditory (non-speech sounds) channels from other domains could apply to the military, we conducted an exploratory survey soliciting opinions from laymen and experts on communicative abilities of a small non-anthropomorphic ground robot. For this study, we used Clearpath Jackal, equipped with a string of LED lights and a speaker to produce nonverbal cues in three different vignettes, and obtained feedback from 16 participants. Our study revealed a number of important issues in nonverbal signal design: importance of context and individual differences, challenge of ambiguity of meaning, and disambiguation of multiple simultaneous messages.

Keywords: Human factors · Robotics · Unmanned systems ·
Military applications

1 Introduction

We live in exciting times, when robots are slowly, but surely moving from the realm of sci-fi, to become an integral part of our society. Unlike the human-like robots of books and movies, the widely-deployed robots of today and the near future are much less anthropomorphic, be it a robot vacuum cleaner like Roomba, a hotel delivery robot like Relay, a quadcopter, or a self-driving car. To successfully integrate into society, these robots need to be able to communicate with people in various roles as customers, teammates or bystanders in order to perform tasks, collaborate, or simply co-exist with people. Transparent communication is a prerequisite to forming accurate mental models of what the robot can and cannot do, and to forming trusting, collaborative interactions. Current solutions focus on use of remote operation via tablets, speech, or direct physical interfaces on the robot itself, such as touch screens or buttons. However, these interfaces are not always feasible (e.g., when network connection is not available, or ambient noise precludes usage of speech), and not always desirable - as in the case of

© Springer Nature Switzerland AG 2020
J. Chen (Ed.): AHFE 2019, AISC 962, pp. 181–193, 2020.
https://doi.org/10.1007/978-3-030-20467-9_17

bystanders, or high-cognitive-load tasks, where traditional interfaces result in information overload. In these cases, the only means of communication left for non-anthropomorphic autonomous robots are nonverbal.

In this paper, we focus on the military robot domain, where radio silence, precluding direct robot control, and high cognitive loads are par for the course. Some robots, like IED detection robots, are already frequently used in the military setting, but the potential for robots that can work even closer with people, at a squad level, are becoming more prevalent [1]. To better understand how the findings in nonverbal communication using visual (lights) and auditory (non-speech sounds) channels from other domains could apply to the military, we conducted an exploratory survey soliciting opinions from laymen and experts on communicative abilities of a small non-anthropomorphic ground robot (Clearpath Jackal), equipped with a string of LED lights along the bottom perimeter, and a small speaker. For the remainder of the paper, we first describe the relevant research that influenced our choice of situations and stimuli, followed by the description of the study and the methods used, concluding with the survey results and a discussion section.

2 Related Work

To understand whether there is a need for non-verbal communication with autonomous systems in the military, we conducted a preliminary investigation first. This investigation was two-fold: interviews with retired military personnel (an Airforce airplane pilot and an Army helicopter pilot), and a short questionnaire given to Soldiers after an experimental field study for a human-robot interaction project. The interviews uncovered a few non-verbal communication techniques currently in use by pilots, such as using tail lights in sequence to maintain formation, landing lights, banking left and right to increase prominence, and others; the pilots also acknowledged the need for unmanned platforms to have similar means of communication. For the field study, we asked the Soldier participants whose task was to control a UGV (Unmanned Ground Vehicle) using hand gestures *how well they could understand what the vehicle was doing*, based on actual experience and hypothetically. The 7-point Likert-style questions were as follows: "How well could you/would you be able to: (1) understand what the vehicle was doing; (2) be able to predict its next step; (3) know if the vehicle understood operators command; and (4) understand when something was going wrong. These questions were posed in 4 different scenarios, for autonomous vs manually driven vehicle, and having or not having speech feedback through a handheld device. As shown in Fig. 1, overall, the Soldiers were unsure of being able to understand UGV behavior *without* explicit communication.

These prelimi-nary findings, com-bined with the increased projected utilization of unman-ned systems, speci-fied as a near-term strategy in the DoD Unmanned Systems Integrated Roadmap [2] assured us of the timeliness of the proposed exploratory research.

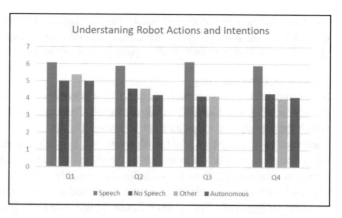

Fig. 1. Without speech feedback from the vehicle, the expectations of being able to understand decreased; they also decreased as the questions became more specific.

The primary goal of the study was to determine whether research findings from an existing body of work in nonverbal communication with non-anthropomorphic platforms can be applicable to the military domain. Therefore, we conducted a literature search in order to determine the specific nonverbal expressions, and select a set of scenarios in which they were proven to be of use (including a wider domain of non-robot devices, such as smart phones [3], electronic triage devices [4], ubiquitous computing [5], and autonomous driving [6]). Based on this literature search, we have identified three separate scenarios/vignettes which could be of potential military relevance: requesting help, indicating directionality, and attracting attention. The remainder of the section describes the related work that informed our survey design and stimuli implementation.

Requesting Help

In a study conducted by Baraka, Paiva, and Veloso [7], the researchers wanted to investigate the use of lights as it related to the robot's state among several scenarios to inform, influence, and cause the human to interact. The scenarios they selected were: waiting for human input, robot's being blocked by an obstacle, and indicating task progress. Specifically, important for our research were the first two situations. After conducting their experiment, it was found that the most effective light for waiting was a light blue color, as they believe cold colors attract attention best. For the blocked situation, red was deemed the most effective as it is seen as demanding.

Another study, done by Cha and Mataric [8], focuses on the most effective way for a non-humanoid robot to request help utilizing lights and sound at varying levels of urgency. Their results indicate that the participants found it better to combine the two types of signals. Sound would tell the participant that the robot needed something, then a blinking light would suggest the level of urgency. In this experiment, yellow represented low, and red represented high urgency.

The last relevant piece of literature was a nonverbal communication literature survey completed by Cha, Kim, Fong, and Mataric [9]. In this paper, the researchers highlight that urgency does not solely rely on light color but could depend on how quickly the light is blinking. For instance, think of a car or smartphone, it will blink red at a relatively high frequency when running out of power.

Indicating Directionality
Szafir and Mutlu [10] worked with a flying robot to determine the best method of communicating directionality while in flight. They designed four behaviors, two of which the participants believed to be effective. The first design had the robot blinking a wide section of a light strip in the direction that it would travel, like a car blinker. The other design was based off human eyes, so the robot had two areas of light lit up in the direction it intended to travel. The color the researchers felt was best was blue as it did not hold connotations such as green for "go" and red for "stop".

Another important finding from Szafir, Mutlu, and Fong [11] is that it is essential to keep a moving robot at a constant velocity. If you do not, it could easily cause the participant anxiety as they would not know where the robot will travel or fear that the robot may run into them.

Attracting Attention
In addition to keeping a robot at a con-stant velocity, it may be important to not have them move in an entirely straight line. This concept was also drawn from Szafir, Mutlu, and Fong [11]. The researchers felt this an important attri-bute because humans cannot walk in a perfectly straight line, so you would not expect or feel comfortable with a robot that could do so.

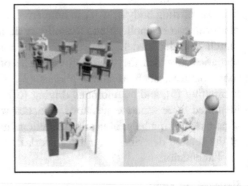

Research done by Takayama, Doo-ley, and Ju [12] was aimed at determining whether people can figure out what their robot is doing based on specific behav-

Fig. 2. PR2 robot scenarios

iors that are similar to humans. Specifically, the robot showed thought before acting and reacted to their success or failure of the given task. Fig. 2 visualizes the four tasks that their robot, PR2, would conduct. These scenarios were: delivering a drink, ushering a person, opening a door, and getting power. From this study they made the important conclusion that to produce the best results of the human maintaining their attention on the robot, the robot should "gaze" at them with lights and stay oriented in their direction.

3   Exploratory Study Design

As this study was exploratory, there were no experimental conditions. A small number of participants were asked to view a non-anthropomorphic robot's nonverbal cues in a series of short vignettes, compare different presentations of cues, and evaluate them with respect to a number of characteristics. After every vignette, the participants answered a few short questions in an online questionnaire, including a number of open-ended questions; they were also encouraged to voice their opinions throughout the study. Each session lasted approximately 20–30 minutes. The scenarios, setup and stimuli/nonverbal cue design were based on both the literature search and cost/availability of the hardware. The latter necessitated that some portions of the vignettes were hypothetical, and the participants had to imagine how the robot would act or nonverbal cues would be presented. The primary nonverbal modalities that we chose were visual (LED lights as a strip around the robot's perimeter) and auditory (nonverbal sounds), supplemented by the robot's movement where appropriate.

Scenario and Setup

In the overall scenario, the robot was presented to the participants as if acting as a part of a human-robot squad, performing a role of a junior squad member and tasked along with an assigned fire team. There were three different vignettes: attracting a teammate's attention, requesting help from a teammate, and indicating directionality (of robots intended movement, as well as location of detected enemy/friendlies). The participants were given a brief introduction to the robot first, and then each of the vignettes was first described, and then robot presented a variety of appropriate nonverbal cues, in sequence. The auditory and visual cues were presented both separately (e.g., lights without sound) and together. The study was setup in a small conference room with overhead lighting and low ambient noise. The participants were sitting in an office chair and were asked to use their imagination to help visualize each of the vignettes. An example setup is shown in Fig. 3, present-

Fig. 3. Study setup: relative positioning of participants and the robot; the experimenter was present in the room, generally slightly behind and to the side of the robot

ing relative positions of participants and the robot, and the average distance between them. In the following vignettes, the robot was fully teleoperated, and the LED lights and nonverbal sounds were selected and turned on/played by the experimenter. We describe stimuli/cue design for each vignette in a later subsection.

In the first vignette, *attracting a teammate's attention* (with the goal of communicating some further information, for example), the robot approached the participant, orienting itself towards him/her. Then, three different colors of static lights were displayed in succession followed by individual sounds being played. The displays were repeated at participants' request, if asked. Once the participant decided which light and sound option were the best for this scenario, a combination of the two were presented.

For this vignette, the cues were designed to be friendly, non-threatening, yet attention-grabbing.

In the second vignette, *requesting help*, the participants were asked to imagine that the robot is in need of help from its teammate – e.g., blocked by an obstacle that it cannot safely navigate around (in the setup, a trash can was placed in front of the robot). The cues for this vignette were designed to express varying degrees of urgency through the choice of color, sound qualities, and repeated exposure (blinking or pulsing for lights, and repeated playbacks for sound). Each of the cues was presented in succession, and repeated if requested.

The last vignette, *indicating directionality*, was split into three separate situations. For all three situations, due to hardware limitations (inability to select LED lights individually), the participants were asked to imagine that only a portion of the LED strip lights up. Static lights without sound were used for all of the situations, primarily to elicit opinions about the colors, and solicit ideas on how to convey the information. The first situation was if the robot was indicating which direction it would move – using static light on the side of the intended turn. Next, we asked participants to imagine that the robot was indicting direction of sensed enemy or friendly forces, where again a static light display was used on the side of the relative location, but of a different color for enemy vs friendly.

Robot Platform and Stimuli/Cue Design

The non-anthropomorphic platform used for this study was a small indoor/outdoor ground robot: Clearpath Jackal, which we equipped with a strip of remote-controlled, non-individually addressable LEDs around the perimeter and a small USB-powered speaker positioned in the center of the chassis (Fig. 4); the Kinect mounted on top of the robot was not used in this study. The lights had the ability to remain a solid color (static), blink (be turned on and off, like a car blinker), or pulse (a gradual brightness change) at several different rates from a variety of colors. Only a subset of available colors was

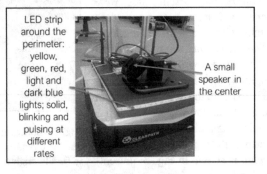

Fig. 4. Jackal equipped with auditory and visual payload

used: green, yellow, red, white, light blue, and dark blue; this was partly due to the literature findings, and partly due to the hardware capabilities, as some colors were not very distinguishable.

Nonverbal cues design space can be defined by two dimensions: *Channel* and *Intrinsicality*, differentiating payload from intrinsic robot features, and visual from auditory communication channels. We focused primarily on those modalities that can be added as a payload to any non-anthropomorphic platform; while robot motion presents a powerful means of nonverbal communication, it is beyond the scope of this study. Table 1 summarizes this design space succinctly.

Table 1. Nonverbal communication design space.

Intrinsicality \ Channel	Visual	Auditory
Robot Motion*	Robot movement with a communicative purpose	Consequential sound (e.g., engine rev-up, breaking, etc.)
Robot Payload	LED lights, laser pointer, projectors	Sound generated to convey urgency, alert, progress, or para lingual cues in speech

Specific choice of color, sound and frequency/rate of blinking/pulsing and sound repetition was informed by the literature search described in Sect. 2, though limited in some cases by the hardware. For the first vignette, attracting attention, the colors presented were green, light blue, and dark blue, and there were three sound options: machine-like/mechanical, human-like (whistle), and cartoonish (movie character – R2D2). The chosen colors and sounds were intended to be friendly, yet attracting attention, and the goal was not to find the best color/sound combination, but rather assess the general feasibility of using visual and auditory nonverbal cues for attracting attention in operationally-relevant tasks. In addition to static colors, both blinking and pulsing lights were also shown, to determine whether that would make the light cues more effective. For the second vignette (requesting help), we were particularly interested in how different colors, sounds and color frequency can convey urgency. Therefore, the stimuli options included displaying red and yellow static lights, blinking lights, and pulsing lights at varying frequencies. The participants were also presented with a two-tone alarm sound (similar to a fog horn) as well as a single-tone alarm sound (similar to an ambulance). Finally, only static lights were used in the third vignette (indicating directionality), with white showing robot's intended movement, red – location of the enemy forces, and green – friendly or neutral forces. Table 2 summarizes our stimuli design choices.

Participants

There was a total of 16 people who volunteered to participate in the study. Of the 16, two were experts in human-computer/robot interaction, and the rest occupied a variety of technical and non-technical positions in our company. All were well-educated and had a basic understanding of computers and technology, though no significant robotics experience (with 2 exceptions). Only two participants had prior active duty experience in the military.

Table 2. Stimuli design.

Stimulus design	Lights		Sound
	Color	**Frequency**	
Attracting attention	Green, Light Blue, Dark Blue	Static	3 sounds: mechanical, whistle, R2D2-like
Request for help	Yellow and red	Blinking vs pulsing; different frequencies	Two-tone vs single-tone; alarm sounds
Indicating directionality			
Robot movement	White	Static	None
Enemy location	Red	Static	None
Friendly location	Green	Static	None

Survey

In addition to note-taking on comments throughout the study, we received participant feedback through an online survey which they answered incrementally after each vignette. Most of the questions were 5-point Likert style, but we also included a few open-ended comment questions. Following the survey, each person was asked to fill out a demographics questionnaire as well. The survey questions were designed to elicit the overall impressions of the robot's nonverbal cues, and feedback on their potential applicability in the military environments. Table 3 summarizes the questions by vignette.

Vignette	Questions	Question type
Attracting Attention	To what extent do these <colors, sounds> express friendliness and draw your attention?	5-point Likert-style, with 1 anchored at "not at all"
	To what extent did each of the options <lights, sounds, combination> draw your attention?	5-point Likert-style, with 1 anchored at "not at all"
	Other suggestions	Open-ended
Requesting Help	To what extent do these <colors, sounds, light patterns, frequency> express urgency?	5-point Likert-style, with 1 anchored at "not at all"
	Do sounds or lights express urgency to a greater extent?	Multiple choice
	Other suggestions	Open-ended
Indicating Direction	To what extent does <white, red, green, respectively> indicate <robot's intended movement, enemy location, friend location>?	5-point Likert-style, with 1 anchored at "not at all"
	Other suggestions	Open-ended
Overall	To what extent were the following appropriate on the robot <volume, brightness, position of lights, position of sound>	5-point Likert-style, with 1 anchored at "not at all"
	Other comments	Open-ended

4   Study Results and Discussion

The overall objective of this exploratory study was two-fold: (a) determine feasibility of using visual and auditory payload-based nonverbal cues by a non-anthropomorphic robot to communicate with human teammate in military-relevant scenarios; and

(b) uncover issues and nuances that can be lost in more formal studies. While the small sample size would not allow us to draw any statistically-backed conclusions, the survey nonetheless provides valuable insights into how nonverbal communication can be used.

Results

Overall, we found that the participants responded positively to the applicability of lights and sounds as a method for the robot to convey information in different types of situations, with the average response to all "appropriateness" questions ranging between 3.6 and 5 (with the exception of the mechanical sound, which was universally disliked as a means of attracting attention in a friendly manner). We present some of the more interesting descriptive statistics below. Figures 5 and 6 display the specific results for two of the vignettes, showing comparative preferences for color, sound, and type of light patterns, as well as differences between HCI experts and laymen. Figure 7 indicates the overall appropriateness for the setup of the robot: above 4 for all categories. On average, it appears that the experts found the stimuli less compelling than laymen as to a robot's ability to request help or indicate direction nonverbally.

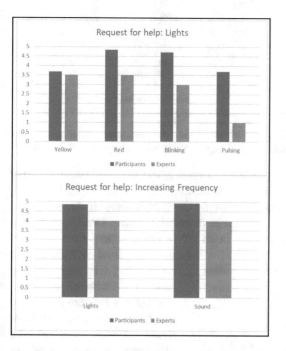

Fig. 5. (top) Red is a better indicator of urgency than yellow, and blinking pattern is perceives as more salient. (bottom) Participants associated increasing frequency with greater *perceived* urgency.

For the first vignette (*attracting attention*) the consensus was that although a green light is friendly, it has additional connotations, such as indicating battery status. Because of this observation, people generally preferred a blue greeting light. Another comment that many participants offered for this scenario was that the sound accompanying the light was too loud. Dependent on the environment around the person and robot, the sound should be adjusted, as should color brightness if you are in a darker or brighter environment.

Next, in *requesting help* vignette (Fig. 5), red was the preferred color by laymen, but the experts found both yellow and red reflecting urgency to a similar extent. Some participants suggested increasing color brightness and sound volume as situations become more urgent. As Fig. 5 indicates, increasing frequency of sound and blinking lights also correlates with increased urgency. Another finding in this scenario was that blinking lights were preferred over pulsing. Some participants commented that pulsing lights were too similar to a heartbeat which might confuse people on what the signal is intending. One participant commented on how important sound is in this instance because you cannot expect the soldiers to always be paying attention to the robot.

Last, in the *indicating directionality* vignette, the laymen found the signals to be more compelling than the experts did, especially for enemy/friend indication. The experts gave us information regarding what certain colors in military settings may indicate. For instance, "friendly" is generally associated with blue ("blue forces"), so it would make more sense to use blue than green. Other people wondered whether the robot could indicate the location of an unknown person, possibly

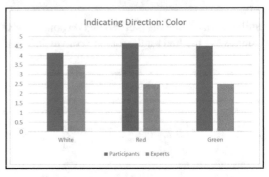

Fig. 6. Laymen found red and green appropriate for indicating enemy vs friend, while experts had reservations.

using the yellow color. Some additional comments included that indicating directionality is rather ambiguous because how would the soldier know if the robot was turning in a more curved way than a straight or diagonal direction. One suggestion to overcome this was to use a moving light in an arc (a chase pattern from the front towards partial side coverage), representing a turn radius. Another participant suggested also using a different light if the robot was backing up instead of turning, similar to a car. An intriguing idea that one of our participants had was utilizing a laser pointer, rather than lights, to show where enemy/friendly forces are. In this case, you could make it a wider light or narrow beam depending on how number of people detected.

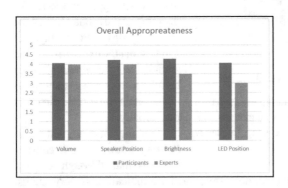

Fig. 7. Laymen found red and green appropriate for indicating enemy vs friend, while experts had reservations.

Some of the major general take-aways participants commented on was the following: they felt there should be an additional strip of lights higher up on the robot so that it would be easier to see. This then would not require the soldier to stare at the ground while walking. On a mission, the robot should not be their priority, instead it should be whatever their objective is. Some participants wondered what if the soldier was color-blind, how would that affect the way the robot utilized lights, if at all.

Discussion

One of the main themes that has emerged from the participants' feedback is the importance of **context** in the design of nonverbal communication cues, especially military context, in our case. One of the biggest challenges we discovered with using sound and lights as means of nonverbal communication was that in tactical situations the robots, just like the Soldiers, should not be flashing lights or making noise. What are alternative communication means in that case? Some ideas our participants had included having the robot link to a headset and produce sound that way, so that only you can hear it, and for lights, using infrared lights to limit its visibility to the enemy, assuming the soldiers are equipped with the necessary technology, such as Night Vision Goggles. Military context (e.g., the need to be stealthy) can be viewed as part of larger *situational* context: for example, attracting attention in urgent vs non-urgent situations. Our participants pointed out that both volume and brightness could vary depending on whether it's a greeting (lower on both) or an urgent request for help (louder and brighter). Other types of context that were evident from the study were: environmental, personal/individual, and platform. The environmental context is rather straightforward: both brightness and volume should be adjusted according to ambient light and noise in order to be salient, yet not overpowering. Ambient light may also need to be considered when selecting color hue – pale colors won't be visible in bright light. To the extent possible, design of nonverbal cues should also take into consideration *individual differences* in human visual and auditory processing; for example, adjustments would be necessary in case of color-blindness. Finally, positioning and size (both length and width) of lights, as well as the types of nonverbal sounds, should depend on the *physical robot platform*. What may be visible on a small robot designed to work in close proximity to humans will be entirely lost on a larger platform used for operation at a distance. Similarly, participants noted that the sounds should match the platform's potential to inflict harm (lower, louder sound for a larger, more dangerous platform, especially during movement).

Another common theme that has emerged is the ***inherent ambiguity*** of both colors and sound types. Different professions and cultures interpret certain sounds and colored lights in a variety of different ways. The ambiguity can even refer to different situational contexts: for example, green-yellow-red is an accepted color progression for denoting go-caution-stop, but in combination with blue, the same color red would mean "hot" as opposed to "stop". Because of this, a universal solution without the proper training is unlikely. That also leaves the question of whether you could train the robots to adapt in specific disciplines or retrain it based on the environment it is in.

Finally, there's a robot architecture design issue needs to be considered. In complex situations, when multiple issues need to be communicated at once, there need to be well-defined arbitration protocols to select appropriate information items, and either display them in succession, or in sequence. For example, in a situation in which the robot was turning, but also detected an enemy, should it only display the more urgent message, and which one would it be, if so?

5 Conclusion

In this paper, we described an exploratory study designed to determine the feasibility of using nonverbal visual and auditory cues for communicating with a non-anthropomorphic robot in military-relevant situations. By primarily relying on existing findings in other domains, we were able to quickly implement several alternative nonverbal cue designs, using a small ground robot, Clearpath Jackal. We then presented them to 16 participants in three different vignettes and solicited their feedback with regards to the cues' appropriateness for the vignettes they were presented with. Our study was largely successful in confirming the overall feasibility of the approach, and helped uncover several considerations for nonverbal signal design, and directions for future work. These considerations include the importance of various types of context (situational, environmental, individual, and platform), the challenge of ambiguity of meaning (varying by culture, profession, training, or even situation), and the need for signal arbitration to disambiguate between multiple simultaneous messages.

References

1. Moshkina, L., Saucer, T., Spinola, M., Crossman, J.: Variable fidelity simulation and replay for unmanned autonomous ground vehicles. To appear in Proc. SAE World Congress Experience (2019)
2. US Department of Defense. Unmanned Systems Integrated Roadmap FY2011-2036. LULU COM (2015)
3. Minsun, K., Lee, J., Lee, H., Kim, S., Jung, H., Han, K.: The color and blink frequency of LED notification lights and smartphone users' urgency perception. In: International Conference on Human-Computer Interaction, pp. 621–625 (2014)
4. Lenert, L., Palmer, D., Chan, T., Rao, R.: An intelligent 802.11 triage tag for medical response to disasters. In: AMIA Annual Symposium Proceedings, vol. 2005, p. 440. American Medical Informatics Association (2005)
5. Tarasewich, P., Campbell, C., Xia, T., Dideles, M.: Evaluation of visual notification cues for ubiquitous computing. In: International Conference on Ubiquitous Computing, pp. 349–366. Springer (2003)
6. Löcken, A., Heuten, W., Boll, S.: Enlightening drivers: A survey on in-vehicle light displays. In: AutomotiveUI 2016 - 8th International Conference on Automotive User Interfaces and Interactive Vehicular Applications, Proceedings, pp. 97–104 (2016)
7. Baraka, K., Paiva, A., Veloso, M.: Expressive lights for revealing mobile service robot state. In: Robot 2015: Second Iberian Robotics Conference, pp. 107–119. Springer, Cham (2016)
8. Cha, E., Matarić, M.: Using nonverbal signals to request help during human-robot collaboration. In: Intelligent Robots and Systems (IROS), 2016 IEEE/RSJ International Conference on, pp. 5070–5076. IEEE (2016)
9. Cha, E., Kim, Y., Fong, T., Mataric, M.J.: A survey of nonverbal signaling methods for non-humanoid robots. Found. Trends Robot. 6(4), 211–323 (2018)
10. Szafir, D., Mutlu, B., Fong, T.: Communicating directionality in flying robots. In: Proceedings of the Tenth Annual ACM/IEEE International Conference on Human-Robot Interaction, pp. 19–26. ACM (2015)

11. Szafir, D., Mutlu, B., Fong, T.: Communication of intent in assistive free flyers. In: Proceedings of the 2014 ACM/IEEE International Conference on Human-Robot Interaction, pp. 358–365. ACM (2014)
12. Takayama, L., Dooley, D., Ju, W.: Expressing thought: improving robot readability with animation principles. In: Human-Robot Interaction (HRI), 2011 6th ACM/IEEE International Conference on, pp. 69–76. IEEE (2011)

Human Interaction with Small
Unmanned Aerial Systems

Spatiotemporal Analysis of "Jello Effect" in Drone Videos

Yang Cai[1]($\boxtimes$), Eric Lam[2]($\boxtimes$), Todd Howlett[2]($\boxtimes$), and Alan Cai[1]($\boxtimes$)

[1] Carnegie Mellon University, 5000 Forbes Avenue, Pittsburgh 15213, PA, USA
ycai@cmu.edu, acail@andrew.cmu.edu
[2] Air Force Research Lab, Rome, NY, USA
{eric.lam, todd.howlett}@us.af.mil

Abstract. A computational method for analyzing the "jello effect" caused by high-frequency vibrations of the camera with a rolling shutter mechanism is demonstrated. We use an optical flow model to analyze the "jello effect" in drone videos. Motion vectors and hue-saturation-value color map are used to visualize the spatiotemporal patterns. Our observations show that poor insulation between the drone structures and the camera creates a persistent "jello effect." An impact of a projectile on the drone can cause a brisk "jello effect" where the magnitude of the shearing, rotational, and translational movements are proportional to the energy of the impact. By scaling the motion vectors in the optical flow, we are able to amplify the movements in the video to observe subtle distortions that are invisible to the naked eye. Our approach combines optical flow and visualization methods for real-time video analytics of drone footage. The study explores inferring physical interactions through non-ideal effects on digital systems. Applications include near real-time drone situational awareness.

Keywords: Jello effect · Vibration · Optical flow · Visualization · Color space · Motion vector · Drone · Video analytics · Dynamics · Cyber-Physical MASINT · Autonomous vehicle · Motion analysis · Computer vision · Multimedia

1 Introduction

Many consumer electronics products such as cell phones, drones, and webcams are equipped with CMOS (Complementary Metal Oxide Semiconductor) image sensors. Compared to the conventional CCD (Charge-Coupled Device) sensors, CMOS imaging sensors are relatively low power consumers, inexpensive, and can be easily integrated into a system-on-chip solution. CMOS sensors typically use rolling shutter for capturing images, which exposes the image row-by-row, instead of grasping all pixels at the same time. This process leads to image distortion. The "jello effect" is a phenomenon that appears when the camera is vibrating at a high frequency. The rolling shutter causes the image to wobble unnaturally. In video footages, the "jello effect" occurs when the camera is not properly isolated from the vibrations of the moving parts, or a projectile object hits the camera system (e.g. drone). Figure 1 shows a frame

© Springer Nature Switzerland AG 2020
J. Chen (Ed.): AHFE 2019, AISC 962, pp. 197–207, 2020.
https://doi.org/10.1007/978-3-030-20467-9_18

of the "jello effect" footage. Motion vectors can describe the phenomenon intuitively, where some motion vectors move toward the right; others move toward the left, creating shearing disturbances in pixels.

Fig. 1. Example of the "jello effect" when the drone was hit by a projectile object. The rolling shutter causes the image to wobble abruptly.

The purpose of this study is to analyze the dynamics of the distortion caused by the camera's rolling shutter to understand the vibration properties, such as spatiotemporal patterns of directions, frequencies, modes, and magnitudes of the vibration. Potentially, the findings can be used for characterizing the vibrations and impact of projectile objects for drones.

2   Related Work

Methods for modeling motion in videos can be traced back to early mobile robotics studies. The prevailing algorithm has been optical flow that tracks visual features between frames. There are two major classes of implementations: Hurn-Schunck model [1] and Lucas-Kanade model [1–5]. The first approach generates intensive optical flow vectors that are appropriate for global modeling. The second approach tracks small patches of pixels in frames that is relatively robust to form sparse local optical flow vectors. Optical flow is capable of modeling spatial and temporal motions from a 2D video. It has been further developed to simultaneously map and locate the camera itself and its surrounding environment in 3D, for example, SLAM (simultaneous localization and mapping) or VO (visual odometry) from single or a stereo pair of cameras [6]. The renaissance of optical flow opens up a broader scope of applications in video analytics in real-time or off-line. It has moved from psychophysical perception modeling to physical and mobile sensing and modeling.

Motion measurement has also been a key technology in video compression algorithms such as MPEG-2, MPEG-4, H.264, and QuickTime [7]. The compression algorithms track the visual features between frames as macro blocks, connected with motion vectors, composing a sequence of compressed predicted frames from an original infra-frame.

Recently, significant studies have been done in measuring vibrations from video footages. By tracking the changes in color and the position of a region of interest, the analytical algorithms can extract vibration modes and frequencies to infer material properties from small, often imperceptible motion in video [8].

Amplification of vibration in motion and color is a breakthrough in video-based motion analytics. The amplification process contains multiple filters such as the spatial filter, temporal filter, and infinite impulse response (IIR) filter. The spatial filter blurs the image to remove the spatial noises. The temporal filter removes the unwanted wave frequencies.

Studies have connected "jello effect" with the rolling shutter mechanism [9]. Ringaby developed the three-dimensional models for rolling shutter movement, including pan and rotation, and associated image distortion [10]. The basic idea is to estimate the camera motion, not only between frames, but also the motion during frame capture. The author uses inter-frame visual feature rather than extracting the motion properties, including the dynamics of the distortion. A non-linear optimization problem is formulated and solved from point correspondences. A set of lab experiments are conducted by the author where all rows in the rolling shutter image are imaged at different times as the motion is known. Each row is transformed to the rectified position. However, the ultimate goal of the study is to correct the image from the distortion rather than spatiotemporal vibration analysis.

3   Representations of Motion Vectors

To extract the motion features, we use Optical Flow to describe the motion, direction, and strength in terms of motion vectors. Optical Flow assumes the brightness distribution on moving objects in a sequence of images is consistent, called "brightness consistency."

$$\frac{dI}{dt} = 0$$

$$\frac{dI(x(t), y(t), t)}{dt} = \frac{\partial I}{\partial x} \cdot \frac{dx}{dt} + \frac{\partial I}{\partial y} \cdot \frac{dy}{dt} + \frac{\partial I}{\partial t}$$

$$= I_x \cdot u + I_y \cdot v + I_t = 0$$

where, I_x, I_y and I_t are spatiotemporal image brightness derivatives. u is the horizontal optical flow (velocity vector). v is the vertical optical flow (velocity vector). We use Horn-Schunck algorithm to minimize the global energy over the image by solving for velocity with iterating over Jaccobi equations. Horn-Schunck algorithm generates a high-density of global optical flow vectors, which is useful to our measurement purpose. We use *grid density* to define the number of motion vectors in a frame. For example, plot a motion vector for every 25 pixels horizontally and vertically respectively. Dynamical visualization of the field of optical flow is a critical component to reveal the changes of the flow patterns overtime, which we call *flow map*. The pseudo code of the process is as following:

```
Get the first frame
Convert the frame to grey-scale
Store as previous_frame
For frame i = 2 to n
        Convert the frame i to grey-scale and store as current_frame
        Calculate the optical flow for previous_frame and current_frame
        Export the flow as motion vectors in a grid
        For y = 0; y < frame.rows; y += grid_density
                For x = 0; x < frame.columns; x += grid_density
                        Point start = Point(x,y)
                        Point flowAtXY = flow(x,y)
                        Point end = Point(x + flowAtXY.x, y + flowAtXY.y)
                        Plot vector (start, end)
                End For
        End For
        Store current_frame as previous_frame
End For
```

In addition to the flow map, we can also visualize the motion vector in the color space of hue, saturation and value (HSV), where hue represents the angle of the vector and value represents the magnitude of length of the vector (magnitude).

$$H_i = \tan^{-1}\left(\frac{y_i}{x_i}\right) \cdot \frac{180}{\pi}$$

$$V_i = \sqrt{x_i^2 + y_i^2} \cdot \frac{1}{M_{max}}$$

where, H_i is hue between 0 and 360, and V_i is value between 0 and 1. x_i and y_i are the optical flow vector's x and y components at point i. M_{max} is the maximum magnitude of the optical flow vectors. The optical flow vector's angle can be naturally mapped to hue in the HSV color system, both in range between 0 and 360°, and the magnitude of the vector can be mapped to a value between 0 and 1. Saturation value for this visualization is constant, so we may set it as 1, the highest value by default. Therefore, we choose HSV color space for mapping the two parameters for simplicity.

4   Experiment Design

We use a DJI drone Mavic Pro™ shooting HD with 1920 × 1080 resolution at 30 fps and 60 fps respectively, on-board GPS, and WiFi/RC remote control. We conduct two experiments: (1) camera vibration induced "jello effect", and (2) projectile object impact-induced "jello effect." In the first experiment, we keep the plastic camera protection cover attached, so the camera is directly attached to the vibrating drone frame, creating a persistent "jello effect." See the left image in Fig. 2. In the second experiment, we use compressed air styrofoam dart shooters to test the impact on the drone's quick vibration. Two types of the shooters were used: one of 8-ft range and another of 80-ft range. We hovered the DJI Mavic Pro™ drone 12 ft above the ground and fired the darts to the bottom of the drone body. The right image in Fig. 2 shows the 80-ft range styrofoam dart we used for experiment.

Fig. 2. Triggers of the "jello effect": the plastic drone camera protector for transportation (left) and the toy projectile missile (right)

5 Persistent "Jello Effect"

In the first experiment, we found that the vibration of the drone's motors triggers horizontal "jello effect," which is periodical and persistent. Figure 3 shows six frames of the video. Figures 4, 5, 6, 7, 8 and 9 show the flow map of the six frames accordingly.

Fig. 3. Six frames of the beach scene from the drone footage that show horizontal "jello effect"

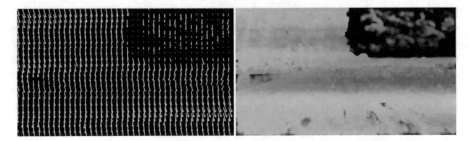

Fig. 4. The flow and color map of the 1st frame of the footage

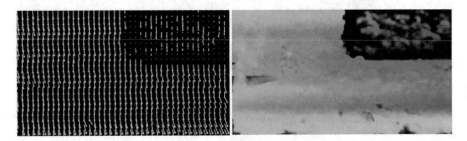

Fig. 5. The flow and color map of the 2nd frame of the footage

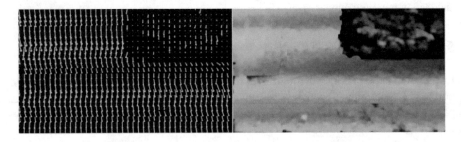

Fig. 6. The flow and color map of the 3rd frame of the footage

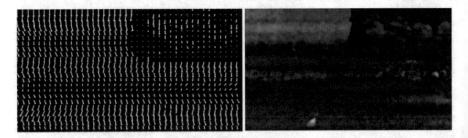

Fig. 7. The flow and color map of the 4th frame of the footage

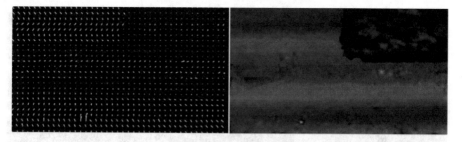

Fig. 8. The flow and color map of the 5th frame of the footage

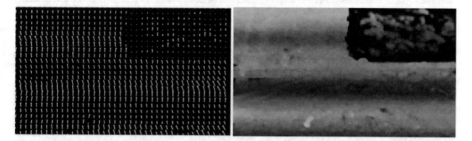

Fig. 9. The flow and color map of the 6th frame of the footage

From Figures 4, 5, 6, 7, 8 and 9, we found that the gap between the peaks of two opposite direction vectors is about 200 pixels. It is 18.5% of the frame height (1080 pixels). Given the frame rate 30 fps in this footage, the detectable vibration frequency is up to 162 Hz. We also found that the area on the up right corner does not change much. This is due to the fact that the clouds in the sky are far away so that there are little visible changes.

6 Brisk "Jello Effect"

When the drone is hit by a projectile dart at a high speed, its video would contain a brisk "jello effect" when the drone tries to regain its control. The wobbling is visible in frames. See Fig. 10.

Figures 11, 12, 13, 14, 15, 16 and 17 show the associated flow maps and color maps of the frames in Fig. 10. The impact occurred at the bottom of the drone. This caused the drone to rotate and displace for a distance. The vibration started from the bottom of the image where it is usually the beginning of the rolling shutter process. From Figs. 11, 12, 13, 14, 15 and 16 we can see the rotation and shearing vectors that caused distortion. The different colors indicate the different direction of the motion vectors. The flight stability mechanism quickly regained control and the "jello effect" disappeared quickly.

Fig. 10. The "jello effect" from the impact of a projectile object

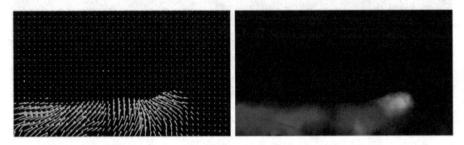

Fig. 11. The flow and color map of the 1st frame of the footage

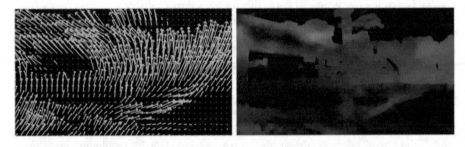

Fig. 12. The flow and color map of the 2nd frame of the footage

Fig. 13. The flow and color map of the 3rd frame of the footage

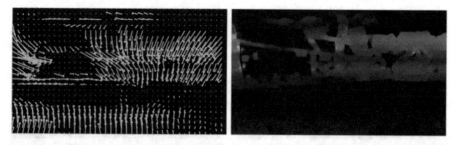

Fig. 14. The flow and color map of the 4th frame of the footage

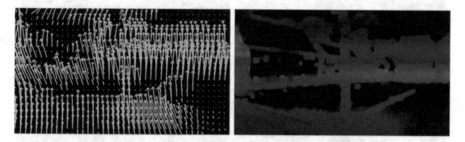

Fig. 15. The flow and color map of the 5th frame of the footage

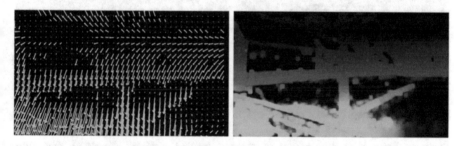

Fig. 16. The flow and color map of the 6th frame of the footage

7   Motion Amplification

Motion vectors in optical flow can be scaled so that subtle movements in the video can be amplified even if they are not visible to the naked eye. For example, the projectile impact of light-weight styrofoam darts (within 10-ft range) to the drone may hardly cause any visible distortion. When we scale up 100 times for the motion vectors, we can see the flow map clearly. Figure 17 shows the tiny rotation and shearing motions in frames, creating subtle "jello effect."

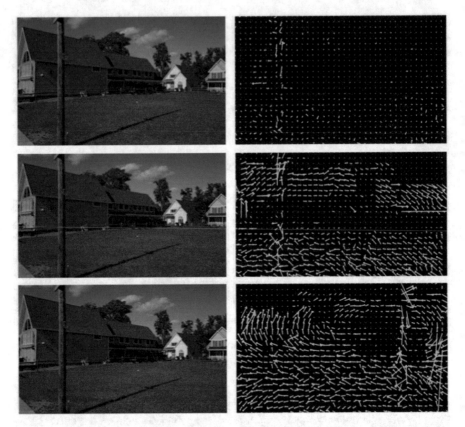

Fig. 17. Flow maps when the drone was hit by a weak dart (the motion vectors are amplified by 100 times)

8 Conclusions

We used the optical flow model to analyze the "jello effect" caused by the rolling shutter process and high-speed vibration of the drone camera. Motion vectors and the Hue-Saturation-Value color map are used to visualize the spatiotemporal patterns. We found that poor insolation between the drone structures and the camera enables persistent "jello effect". On the other hand, an impact of a projectile object on the drone can cause brisk "jello effect" where the magnitude of the shearing, rotational, and translational movements are proportional to the energy of the impact. By scaling the motion vectors in the optical flow, we are able to amplify the movements in the video to observe subtle distortions that are invisible to the naked eye. Our approach combines optical flow and visualization methods for real-time video analytics of drone footages. The study explores inferring physical interactions through non-ideal effects on digital systems and provides an example of an emerging sub-discipline of Measurements and Signature Intelligence (MASINT) using cyber-physical systems. Cyber-Physical MASINT is an area being developed by the US Air Force involving phenomena transmitted through cyber-physical devices and the interconnected data networks to infer information through effects such as digital noise, bit errors, or latencies.

9 Discussions

Optical flow is normally sensitive to the ego motion and the distance of objects in front of the camera. In order to obtain a more accurate estimation of the vibration, it is desirable to measure the distance to the objects in the frame. This can be implemented with Video Odometer (VO) or Simultaneous Location and Mapping (SLAM) algorithms for single camera. With stereo camera or depth sensors, the estimation can be further improved. The camera's motion characteristics (translation and rotation) can also be extracted from the video with the algorithms above. Furthermore, the local distortion caused by the rolling shutter process can be subtracted from the estimated global motion of the camera to reduce the motion noise.

Acknowledgement. This study is in part supported by Visiting Faculty Research Program (VFRP) and the Cyber-Physical Information and Intelligence (CYPHIN2) program at Air Force Research Lab (AFRL) at Rome, NY, 2017.

References

1. Horn, B.K.P., Schunck, B.G.: Determining optical flow. Artif. Intell. **17**, 185–203 (1981). Manuscript available on MIT server
2. Mathworks, Video and Image Processing User Manual (2010)
3. Barron, J., Fleet, D., Beauchemin, S.: Performance of optical flow techniques. Int. J. Comput. Vis. **12**, 43–77 (1994)
4. McCane, B., Novins, K., Crannitch, D., Galvin, B.: On benchmarking optical flow. Comput. Vis. Image Underst. **84**(1), 126–143 (2001)
5. Sheridan, J., Ballagas, R., Rohs, M.: Mobile phones as pointing devices. In: PERMID Workshop (2005)
6. SLAM. https://en.wikipedia.org/wiki/Simultaneous_localization_and_mapping
7. Mandal, M.K.: Digital video compression techniques. In: Multimedia Signals and Systems Springer (2003)
8. Davis, A., et al.: Visual vibrometry: estimating material properties from small motions in video. In: CVPR (2015)
9. https://en.wikipedia.org/wiki/Rolling_shutter
10. Ringaby, E.: Geometric Computer Vision for Rolling-Shutter and Push-broom Sensors. Linköping University, Department of Electrical Engineering, Sweden. Thesis No. 1535, June 2012

Development of a Concept of Operations for Autonomous Systems

Antti Väätänen[✉], Jari Laarni, and Marko Höyhtyä

VTT Technical Research Centre of Finland Ltd,
Vuorimiehentie 3, P.O. Box 1000, 02044 Espoo, Finland
{Antti.Vaatanen,Jari.Laarni,Marko.Hoyhtya}@vtt.fi

Abstract. The paper will present the development of a general concept of operations for autonomous systems and robot solutions. A Concept of Operations (ConOps) of a system is a high-level description of how the elements of the system and its environment communicate and collaborate in order to achieve the stated system goals. The ConOps utilizes the capabilities of autonomous systems, different unmanned vehicles and robots and human operators as a joint cognitive system where the tasks of human operators and robots complement each other. It will include planning the mission, setting up the autonomous system operations and possible robots, monitoring the progress of the operation, reacting and adapting to the intermediate results, reacting to unexpected events and finally completing the mission. The general ConOps can be considered as a template, which can be tailored for the specifics of different use cases.

Keywords: Concept of Operations · Autonomous systems · Robots · Human factors

1 Introduction

ConOps of a particular system is a high-level document describing how the elements of the system and its environment interact in order to achieve the proposed system goals. Typically, a ConOps consists of documents, illustrations, and other artefacts describing the characteristics and intended usage of the system or system of systems from the viewpoints of their users. We propose the ConOps as a knowledge sharing artefact in the design and evaluation of a complete autonomous system. We also propose that the specific ConOps should be maintained and updated throughout the system life-cycle as an overview description of system goals and policies.

The ConOps for autonomous systems should consider the capabilities of different system features and human operators from the perspective of a joint cognitive system where the tasks of human operators and system elements complement each other. It will include planning the tasks and operations, setting up the system components, e.g., robots, monitoring the progress of the operation, reacting and adapting to the intermediate results, reacting to unexpected events, and finally completing the mission. Levels of automation for each phase of a mission is also determined in the ConOps document. The general ConOps will be complemented with the specifics of different use cases.

© Springer Nature Switzerland AG 2020
J. Chen (Ed.): AHFE 2019, AISC 962, pp. 208–216, 2020.
https://doi.org/10.1007/978-3-030-20467-9_19

The ConOps for autonomous systems alternatives can be developed through three main phases: First, in the familiarization phase, data on early solutions, developmental needs, and operational scenarios should be investigated; second, in the data collection and analysis phase, user and performance requirements will be collected and analyzed, and the first versions of the ConOps will be drafted; and third, in the commenting phase, the complemented and final versions of the ConOps are prepared, delivered and discussed with the end-users. Typically, ConOps drafts are developed through a series of workshops, topics of some of which are shown below (Fig. 1)

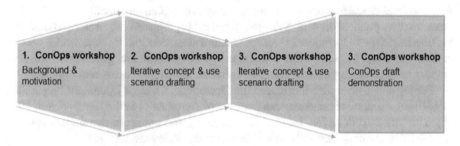

Fig. 1. ConOps development process (modified from [12])

Overall, the present approach to ConOps development provides an essential starting point for the more detailed design of autonomous systems, for example, by enabling to better understand the motivations and needs of different user groups.

2   Concept of Operations Background

A ConOps can be considered as a special kind design artefact playing an important role in the requirements specification activity at the early stages of the design process, and as an overview description of overall goals and policies throughout the system life-cycle. In fact, the ConOps document should support communication and collaboration and provide a basis for discussions about the target system at all stages of the development process [1].

A ConOps of a system is a high-level description of how the elements of the system and its environment communicate and collaborate in order to achieve the stated system goals. Typically, a ConOps consists of documents, illustrations and other artefacts describing the characteristics and intended usage of proposed and existing systems from the viewpoints of their users. Even though a ConOps is developed at the early stages of the design process to promote a common understanding of design goals, it should be maintained throughout the lifecycle of the system by updating it whenever changes are made to the system. The aim is that the ConOps outcomes would also function as a living document that could be updated throughout the system development process [1].

The Concept of Operations term was introduced first time by Fairley and Thayer after mid 90's [2]. They described the ConOps document utilisation in the design of software-based system, and they illustrated ConOps development process and the benefits of ConOps documents. According to Fairley and Thayer, the US military standards were the first documents introducing 'Operational Concept Document' term, which was used before the ConOps term and approach became common.

US military standard DoD-STD-2167 [3] for Joint Logistics Commanders includes a specific "Operational Concept Document" (OCD) section which describes the purpose of the system, its operating environments, and the role of a specific system of systems. This joint regulation contained a Data Item Description (DID) that aims to describe the mission of the system, its operational and support environments, and the functions and characteristics of the computer system within an overall system [3]. Ten years after its publication MIL-STD-498 standard [4] replaced the earlier standard and included a detailed Operational Concept Document description. Its commercial version is IEEE Standard 1498 [5] which aims to merge commercial and government software development requirements and consider software life-cycle process requirements [2].

Fairley and Thayer also introduced the Operational Concept Document Preparation Guidelines standard of the American Institute of Aeronautics and Astronautics in 1992 [6]. This standard describes information gathering and operation definition steps and required documenting methods, and it includes the following topics:

1. Scope, incl. system identification, purpose, and overview;
2. Referenced documents;
3. User-centred operational description, incl. mission accomplishments, strategies, tactics, policies and constraints;
4. Users and their characteristics;
5. Operational needs, incl. mission and personnel needs;
6. System overview;
7. Operational environment;
8. Support environment;
9. Representative operational scenarios, incl. detailed sequences of user, system, and environmental events.

The characteristics of a ConOps document were first defined in IEEE Standard 1362 published in 1998 [7]. This user-oriented document describes a system's operational characteristics from the end user's viewpoint. It is used to communicate overall quantitative and qualitative system characteristics among the user, buyer, developer and other stakeholders. Systems and projects are sometimes parts of larger entities. The ConOps for a specific subsystem may be a separate document, or it may be merged into the main system level ConOps document.

In 2002, International Standard Organization (ISO) published the standard "ISO 14711:2002 Space systems – Unmanned mission operations concepts – Guidelines for defining and assessing concepts products" [8]. The standard specifies areas and products which are relevant in developing operational concepts for space systems missions. It is designed to support experts in industries, government agencies and scientific domains.

ConOps is a central element in ISO/IEC/IEEE 29148 standard concerning requirements specification in the domains of systems and software engineering [9]. This standard addresses ConOps from two perspectives. First, a business level approach focuses on the leadership's intended way of operating the organisation. This approach is grounded on the fact that the ConOps document is frequently embodied in long-range strategic plans and annual operational plans. It thus aims to guide the characteristics of the future business and systems, enable the design project to understand its essential design drivers, and the designers to realize the end users' and other stakeholders' requirements. Second, a more focused "system operational concept" (OpsCon) document describes what the system being designed will do and why. Similarities exist between ConOps and OpsCon, but their main scopes are different.

Tommila et al. [10] and Laarni et al. [11] summarized the main characteristics of a ConOps as follows. The ConOps:

- Is linked with a specific system;
- Can be defined at various levels of hierarchy;
- Illustrates the main elements, functions, states and properties of a particular system and its environment;
- Includes high-level goals, constraints and user requirements;
- Communicates and analyses system goals and constraints;
- Can be understood by all stakeholders;
- Is general enough;
- Illustrates the interplay of system elements in system operation and maintenance;
- Includes a set of information in the system model;
- Is maintained as an overall description of the system and its functioning;
- Includes and integrates both technical and human factors information;
- Provides a basis for the determination of the requirements for the proposed system.

ConOps could be used as a knowledge sharing object that promotes communication and collaboration and integration of design activities in the development of complex systems or system of systems [11]. The level of detail should be optimal so that the description is accurate enough to fully explain how the proposed system is planned to operate. On the other hand, it should be underspecified enough to promote discussions among stakeholders coming from different domains.

ConOps documents are developed and deployed in many domains such as traffic control, space exploration, financial services and different industries such as nuclear power, pharmaceutical and medical [11]. ConOps documents come in different formats, and the main division can be made between user-specific and system-specific ConOps. Some ConOps documents describe the higher-level activities of a set of stakeholders; some documents describe the characteristics of a system from an operational point of view.

2.1 Example Case

Laarni et al. [12] have recently developed a ConOps for autonomous and semiautonomous robot swarms in operational scenarios related to coast guarding by the navy, air surveillance missions and support for the urban troops' operations. The study collected data with literature reviews, questionnaires, cognitive task analysis, and workshops and interviews with representatives of different military branches.

In their study the ConOps for autonomous and semi-autonomous robotic swarms consists of the following topics [12]:

- Overall goals and constraints of the robotic system;
- Characteristics of the environment;
- Main system elements and functions;
- Operational states and modes;
- Operational scenarios;
- Performance requirements;
- System advantages and disadvantages;
- Human-automation interaction concept.

Laarni et al. [12] listed six different management modes according to the complexity of robot swarm operation. In the first and simplest level, the operator assigns a task to one type of swarm and monitors its progress. In the most complicated level, several operators manage swarms of different classes and human teams and vehicles of several units are assigned to co-operate with the robot swarms.

2.2   Concept of Operations Development

The ConOps structure is composed of three main actors: autonomous system, human operators, and other stakeholders. *Autonomous system* can be, e.g., remotely operated or autonomous vessels or Unmanned Aerial Vehicles (UAV) that co-operate with *human operators* in an operation centre. *Other stakeholders* are professionals or officials that are involved in the operation of autonomous system activities.

Based on the identification of the ConOps structure, the initial ConOps diagram can be defined. Figure 2 illustrates the ConOps structure where human operators play a central role and co-operate with other actors such as other stakeholders and the autonomous system. The operating environment consists of climate conditions, landscapes and societal, legislation and regulation issues. For example, International Regulations for Preventing Collisions at Sea (COLREGs) and international frequency regulations for satellite and terrestrial systems need to be considered in autonomous ships domain to enable safe and reliable operations.

The autonomous system in ConOps includes level of autonomy definition, vehicle or device classification, operating modes, operational states and user interfaces for supervising system status and task progress. Operators are addressed when defining topics such as use cases, user requirements, tasks and goals, and operational scenarios. Figure 2 also demonstrates how these different actors and the operating environment affect each other, and what kind of information is needed and shared.

Various knowledge elicitation techniques can be used in gathering use case specific information. We have mainly used critical incident interviews, in which representative events are analyzed by identifying actions, decisions and communications related to the incident. Typically, interviewers prepare use case-specific slide show presentations in order to illustrate use case-examples to the participants. The use case-specific slide presentations may include the following topics:

- Starting point of the use case;
- Definition of participants' role in each use case;

- How the abnormal situations and objects are detected in daily work;
- Communication: how information is received and shared;
- Decision making what kind of information is needed for decision making and how autonomous system could help in processes;
- Duration of missions from the beginning to the end;
- Technology: what kind of technology would be useful and how it could be utilized, e.g., using tablet devices and smart TVs as a user interface.

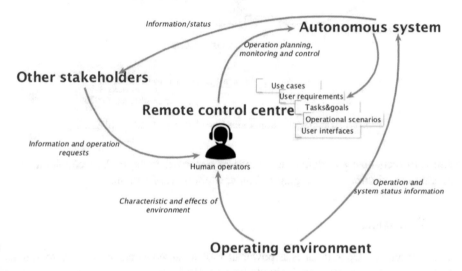

Fig. 2. Initial ConOps diagram

The interviews can provide information about the autonomous system missions, their special challenges and requirements for new technical solutions. Based on the interviews results, the initial ConOps diagrams can be updated. The diagrams aim to illustrate essential elements of different use cases and to define human operator roles and information sharing needs and characteristics. Every diagram includes four main areas:

1. Other stakeholders section lists other officials and their co-operation needs related to the system;
2. Autonomous system describes essential features of the system and what requirements should be considered related to the system;
3. Human operators related information shows operator roles and how they co-operate with other stakeholders and with the autonomous system;
4. Environment section lists use case related environmental characteristics.

The Fig. 3 presents one example of a modified and updated ConOps diagram for an autonomous vessel system. The diagram was developed in a project which aimed to define information requirements for shore control center work. The project studied and identified new work roles for the remote operation of unmanned ships. In addition, it

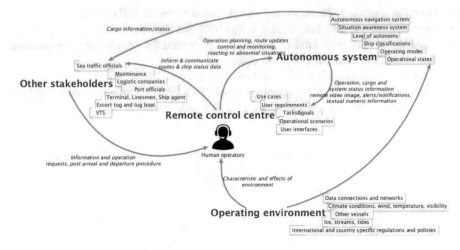

Fig. 3. Example of autonomous vessel system ConOps diagram

aimed to recognize and define new kinds of expertise, tasks, tools, technologies and operation environments emerging in this new era of vessel traffic.

3   Discussion

The Concept of Operations is a powerful tool to demonstrate how an autonomous system should work and how it should be operated in different domains. The ConOps diagram for autonomous vessels lists possible other stakeholders that may affect mission accomplishments and should, thus, be considered when planning boat trips and port activities. The autonomous vessel system is composed of information sharing technologies and technologies required in ships. Human operators will monitor the progress of an operation, and they can react to abnormal conditions and make route replanning if needed. Regarding the level of automation, the ship can be fully autonomous in normal conditions and in predefined routes, but in some situations an operator's intervention (e.g., response confirmation) is needed. On the other hand, abnormal situations may require continuous operator interventions, replanning, and active communication with the other stakeholders e.g., sea traffic and port officials.

As an example, the ConOps diagram can be also used to assist the design of the communication system capable of supporting vessel system operations. Some examples are: (1) seamless logistics and related messaging between vessels, ports, and logistics companies; (2) situational awareness techniques needed and implementation of information delivery from the vessel to the remote operator; (3) implementation of a ship control system (i.e., is the control direct and continuous or does the ship make most of the decisions) [13]. A mature ConOps document illustrated by diagrams can assist the autonomous system designers in building the big picture and promote communication and collaboration.

The initial ConOps can be used to evaluate the autonomous system through the end user interviews, and the use case-specific ConOps diagrams can demonstrate the interplay between different actor roles and the system in detailed level. However, it is still somewhat open, how the ConOps will be deployed in the design of autonomous systems.

One of the main functions of a ConOps is to benefit the development of a system architecture, and specifically, the development of a control concept [11, 12] or the interaction concept [14] for the supervision of the autonomous system. These descriptions characterize operator tasks, their interactions with the technical system and communications with other actors in the domain. As proposed, control and interaction concepts can be used, e.g., in detailed modeling of workload and situation awareness [14].

3.1   Lessons Learned

Some lessons have been learned during the development of the ConOps:

- Enough effort should be allocated to the development of a ConOps in order to achieve detailed insights and understanding about the domain;
- Enrichment of the ConOps by 3D visualizations and simulations may facilitate discussions with different stakeholders;
- ConOps should play an important role in continuous iterative design of complex systems. In order to fulfill the promises, continuous dialogue should be established between ConOps developers and system designers. Otherwise the ConOps would play only a marginal role in system design;
- In order to make the ConOps a valuable knowledge sharing tool throughout the life-cycle of the autonomous system, it should be updated and maintained regularly;
- Systematic procedures are needed for the specification, implementation and maintenance of ConOps artefacts.

3.2   Future Work

The ConOps approach will be used in different projects related to autonomous system design. The main aim is to further develop our ConOps development practices to be better able to tackle integration and communication challenges pertinent to systems engineering in general and human-systems integration specifically. For example, we aim to have better understand how use case specific ConOps results and requirements should be communicated to system developers and other stakeholders who may use different terms and vocabulary related to systems.

References

1. Jost, A.C.: ConOps: the cryptex to operational mission success, crosstalk. J. Defence Softw. Eng. **20**, 13–16 (2007)
2. Fairley, R.E., Thayer, R.H.: The concept of operations: the bridge from operational requirements to technical specifications. Ann. Softw. Eng. **3**, 417–432 (1997)

3. US DoD: Operational Concept Document (OCD), DI-ECRS-8 × 25, DoD-STD-2167, US Department of Defense (1985)
4. US DoD: Software Development and Documentation Standard, MIL-STD-498, US Department of Defense (1995)
5. IEEE Std 1498–1995: IEEE Standard for Acquirer-Supplier Agreement for Software Life Cycle Process, Version D2, The Institute of Electrical and Electronic Engineerings, Inc. (1995)
6. ANSI/AIAA G-043-1992: Guide for the Preparation of Operational Concept Documents. American Institute of Aeronautics and Astronautics (AIAA) (1992)
7. IEEE Std 1362-1998: IEEE Guide for Information Technology—System Definition— Concept of Operations (ConOps) Document. The Institute of Electrical and Electronics Engineers, Inc., New York (1998)
8. ISO 14711: Space Systems—Unmanned Mission Operations Concepts—Guidelines (2002)
9. ISO/IEC 29148: Systems and Software Engineering—Life Cycle Processes—Requirements Engineering. Final Draft International Standard (2011)
10. Tommila, T., Laarni, J., Savioja, P.: SAREMAN Project. Concept of Operations (ConOps) in the Design of Nuclear Power Plant Instrumentation & Control Systems. VTT, Espoo (2013)
11. Laarni J, Koskinen H., Väätänen, A.: Concept of operations as a boundary object for knowledge sharing in the design of robotic swarms. Submitted to Cognition, Technology & Work (2019)
12. Laarni, J., Koskinen, H., Väätänen, A.: Concept of operations development for autonomous and semi-autonomous swarm of robotic vehicles. In: Proceedings of the Companion of the 2017 ACM/IEEE International Conference on Human-Robot Interaction, pp. 179–180 (2017)
13. Höyhtyä, M., Huusko, J., Kiviranta, M., Solberg, K., Rokka, J.: Connectivity for autonomous ships: architecture, use cases, and research challenges. In: The 8th International Conference on ICT Convergence, pp. 345–350. IEEE (2017)
14. Colombi, J.M., Miller, M.E., Schneider, M., McGrogan, J., Long, D.S., Plaga, J.: Predictive Mental Workload Modeling for Semiautonomous System Design: Implications for Systems of Systems. Syst. Eng. **15**, 448–460 (2012)

Human–Drone Interaction: Virtues and Vices in Systemic Perspective

Olga Burukina[1,2] and Inna Khavanova[2,3(✉)]

[1] The University of Vaasa, Wolffintie 34, 65200 Vaasa, Finland
obur@mail.ru
[2] The Institute of Legislation and Comparative Law under the Government
of the Russian Federation, B. Cheremushkinskaya Street 34,
117218 Moscow, Russia
[3] The Financial University under the Government of the Russian Federation,
Leningradsky Prospect 49, 125993 Moscow, Russia
ahavanov@mail.ru

Abstract. The emergence and use of advanced technologies in today's commerce has gradually grown into habitual practice, and the introduction of more modern weapons including UAVs to military operations is hardly a new challenge in the history of armed conflicts. The interim research results concerning attitudes to drone usage have highlighted a number of contradictions in national and international law and policies and revealed a certain inconsistency in the respondents' attitudes partially caused by the different width of Overton windows devoted to drone expansion in the two countries, as well as by the use of the socio-cognitive tools currently changing the national attitudes and value systems as part of the national mentalities. The research has highlighted a number of contradictions that proved to be more profession-specific, age and gender-specific.

Keywords: UAVs · Commercial use · Warfare · Human factors ·
Overton windows · International legislation · National regulations

1 Introduction

Twenty-five years ago nobody could imagine that UAVs or drones would enter our lives so tightly. Of course, they are still far from the popularity of mobile phones, but there is no doubt – a real technological revolution is taking place right before our eyes: small and large, flying and crawling, radio-controlled and practically autonomous – all sorts of drones have been entering and influencing human lives.

Today drones have found application in many areas: filmmakers shoot videos from a bird's-eye view, emergency services investigate dangerous terrains, online stores have been gradually replacing couriers with drones, and drone competitions spreading all over the world – e.g. on 28 July 2018 more than 3,000 guests attended Drone Racing League event at BMW Welt, with the entire Drone Racing League series broadcasted every Thursday from 13 September 2018 until mid-December, and the

© Springer Nature Switzerland AG 2020
J. Chen (Ed.): AHFE 2019, AISC 962, pp. 217–229, 2020.
https://doi.org/10.1007/978-3-030-20467-9_20

winner of the race competing against the other 2018 DRL Allianz World Championship Season winners in their final event in Saudi Arabia [1].

Drones revolutionized the nature of both today's commerce and war, becoming one of the most applicable and successful military achievements in the modern history of armaments. As can be seen from the results of the policy of using UAVs by the armed forces of the United States, Israel, France and other countries, the use of drones in the modern world enlarges entrepreneurs' possibilities and buyers' opportunities, and simultaneously increases the risks of respecting the human rights of civilians facing war conflicts, their right to life and safety in particular. However, the practice of both the modern commerce and warfare has changed due to technological innovations, in particular the expanding use and further development of the technology of unmanned aerial vehicles, or drones.

As technology evolves and the use of drones increases, the international legal system should intensify work aimed at restricting the use of these and similar technologies in accordance with human rights obligations, particularly in the framework of International Human Rights Law (IHRL) and International Humanitarian Law (IHL). Yet, national legislations as well as International Law seem to have been far from correlating each other well enough thus expanding Overton windows and perplexing experienced lawyers.

2   Rise of the Drones

By 2025 the development of UAVs can create 100,000 jobs for the US economy and give an economic effect of $82 billion. The research service of Business Insider predicts annual growth rates of 19% in this industry between 2015 and 2020, with the growth of the military expenditure on drones by 5% only. Amazon, Just Eat, Flytrex, UPS, DHL have been practicing cargo delivery by drones, reducing delivery time drastically. In 2017 Dubai tested a Volocopter drone for passenger transportation.

2.1   Commercial Use

The global market for commercial use of unmanned aerial vehicles has exceeded $127 billion, according to the results of a study by PwC. Currently, the most promising industry for the introduction of drone is infrastructure, with the market valued at $45.2 bn, next come agriculture ($32.4 bn) and transport ($13 bn) [2]. However, even here drones raise enormous privacy concerns as they can be easily abused. Therefore, before acquiring drones, two threshold questions should be asked – (1) whether the local community really needs them and whether it has rigid safeguards and accountability mechanisms in place, so that no one uses drones for warrantless mass surveillance.

2.2   Warfare

The increased use of drones for civilian applications has presented many countries with regulatory challenges. Such challenges include the need to ensure that drones are operated safely, without harming public and national security [3]. However, some researchers indicate that UAVs or drones contributed to the robotization of war. Although they do not belong to the classic robots, since they do not reproduce human activity, experts often rank them as robotic systems. The use of UAVs in various conflicts (Iraq, Afghanistan, Yemen, Somalia, etc.) showed the advantages of the technologies, especially in confronting terrorists.

Yet, the authors share the opinion of Ahmad Qureshi stating that in order to understand the legality of the use of drone strikes, it is relevant to note that Article 2(4) of the UN Charter prohibits the use of force and that Article 51 is the only exemption against such a use of force under the *jus cogens* principles of self-defense under customary international law. Drone strikes, and thereby target killing constitutes an act of war and the use of force can only be justified as self-defence in an actual armed conflict. < ... > Furthermore, even if the drone strikes are legal, do not constitute an act of war against any state, and are conducted with the consent of host states, they must follow certain humanitarian law principles [4].

According to V.V. Karyakin, the use of drones "is not yet governed by international humanitarian law, the rules of warfare and the protection of civilians" [5].

3   The Use and Abuse of Drones

Drones are the first cast of the key that opens the Pandora's Box, and if the stakeholders do not take action today, tomorrow the uncontrolled contents of this box, breaking free from it, can destroy humanity. Today is high time to take the necessary steps.

3.1   Both Parties Easily Abused

The military doctrine of the Russian Federation refers the massive use of unmanned aerial and autonomous naval vehicles, guided robotic models of weapons and military equipment to the typical features of modern military conflicts [6]. The US Department of Defense in Unmanned Systems Integrated Roadmap FY2017-2042 states, "Advances in autonomy and robotics have the potential to revolutionize warfighting concepts as a significant force multiplier" [7].

Besides, the new technology now allows perpetrators to stalk and harass victims, as well as avoid restrictions imposed by restraining orders [8]. Yet, in the absence of sound regulations, both parties (not only the military and civilians or criminals and victims) but *humans* and *robotized self-learning items* will be increasingly abused.

3.2   Human Responsibility

Human responsibility for drone use is three-fold. First, it is strategic – people (both the decision makers and the general public representatives as in the future everyone will be concerned) should work out a global strategy for UAVs and other robotized systems development taking into account the need to preserve the present alignment of forces – humans' rule, not robots. Second, it is humanity's self-preservation – not yet the standoff against robots but primarily the protection against the privileged ones making decisions to use drones and discarding possible poorly justified civilian casualties, with drone operators to face serious consequences for their wrong decisions. Third is the problem of drone privacy as citizens find themselves unprotected from drone intrusion.

4   Research Outline

The undertaken pilot research was aimed at both identifying the ongoing processes in the national legislations of the United States and Russia, which seem to have been experiencing a new round of heightened tension in international relations, and revealing the attitudes of future decision makers – today's Russian university students and their supporters and educators – Russian academics and university managers. The survey was supposed to be conducted simultaneously in Russian and US universities, but the American colleague addressed did not support the idea or even replied to the request.

The research methodology is based upon comparative and systemic approaches to the US and Russian national legal systems in their application to drone usage in such drastically contradicting spheres as commerce and war conflicts and includes analysis of national juridical norms and legal practices, as well as a pilot survey aimed at revealing the attitudes of US and Russian university students to the achievements, failures and prospects of drone application in national and international commerce, as well as the past and ongoing military conflicts, taking into account the previously unseen economic and humanitarian benefits yielded by drone usage, along with the virtualization of drone operators' responsibilities for any possible wrongs.

The research consisted of two parts – one was based on thorough investigation of the current US and Russian legislation and legal discussions, both national and international; while the other was a pilot qualitative research based on a multiple-choice questionnaire.

4.1   Research Goal and Hypotheses

The undertaken pilot research was aimed at both identifying the ongoing processes in the national legislations of the United States and Russia, which seem to have been experiencing a new round of heightened tension in international relations, and revealing the attitudes of the future stakeholders and decision makers as regards to the future development of the Russian Federation – those of todays' students of pedagogy, economics and management.

The survey based on two hypotheses –

Hypothesis 1 – students (people aged under 25) feel freer with innovations and are less cautious concerning their use (UAVs included).
Hypothesis 2 – students (people aged under 25) are more sensitive to decisions taken by single persons (even experts) and not so trustful when it comes to issues that might concern them directly.

The research analysis of the Russian and US legislation and legal and political discourse have identified five challenges facing the growing use of UAVs/drones in the world as both a prominent logistics and warfare tool:

Challenge 1 – Legal: the diversity and inconsistency of national and international legislation have blurred the legal basis for UAVs application, which provides ground for all sorts of improper use and abuse of drones within military conflicts and beyond them.

Challenge 2 – Economic: the economic motives aimed at benefits (profits included) from using UAVs instead of soldiers (who are evidently more vulnerable and quite costly taking into account their training, equipment, supplies, transportation, accommodation, salaries, insurance, hospital treating, pensions, etc.) can surpass and prevail in decisions taken by core stakeholders of leading economies.

Challenge 3 – Political: with UAVs opening new prospects and opportunities in armed conflicts, as well as in all sorts of political conflicts and collisions as both a new means and method of their resolution, they simultaneously turn into both an effective and efficient means of strong political influence, with no country in the world feeling safe enough.

Challenge 4 – Social: practically every society in the world will face a dilemma of looking down on the possibility of unlawful and unjustified civilian casualties in other countries or looking up hoping that drones will never be used to harm them, their families and friends.

Challenge 5 – Organisational: with strong political lobbies, great profits and economic interests of drive manufacturers and the governments of leading economies, the international legislation regulating drones production and application will be kept unchanged as long as possible and organizational difficulties will be used to avoid possible fast solution of the current legal turmoil in regulating the widening use of drones in all spheres including targeted killings inside and outside military conflicts.

4.2 Research Method and Processes

The research methodology was based upon comparative and systemic approaches to the US and Russian national legal systems in their application to drone usage in such drastically contradicting spheres as commerce and war conflicts and includes analysis of national juridical norms and legal practices, as well as a pilot survey aimed at revealing the attitudes of Russian university students, academics and managers to the achievements, failures and prospects of drone application in national and international commerce, as well as in the past and ongoing military conflicts, taking into account the previously unseen economic and humanitarian benefits produced by drone usage, along with the virtualization of drone operators' responsibilities for any possible wrongs.

The designed questionnaire contained 3 background questions and 13 thematic multiple-choice questions intended to identify the general attitude of Russian undergraduate and graduate students and university employees to the drone's employment both in civil and military operations, both inside and outside armed conflicts, as well as opportunities and risks following their predicted and unpredicted potentials.

The participation in the survey was voluntary, which was a delimitation of the pilot research, corresponding to its goal and conditions: the questionnaire was printed out on paper and distributed among students and employees to be collected 20 min later.

4.3 Respondents

Aimed at a pilot research, the sample was limited to 100 respondents – 75 Russian undergraduate and graduate students from one Russian university – Lipetsk State Pedagogical University – and 25 Russian academics and managers from the same university. The random sampling covered over 70 third and fourth-year Bachelor students and first-year Master students, with 57 of them providing valid responses in the questionnaires, and 25 employees of the same university, with 22 of them providing valid responses. Both groups of the respondents took part in the survey in similar research situations. The students returned 71 questionnaires, with 14 of them containing insufficient information for further analysis and the employees returned all the 25 questionnaires, with three of them lacking part of the required information to the extent necessary for further analysis.

The described situation has revealed a problem of either a lack of interest to the research topic or insufficient time length due to the novelty of the topic to both groups of the respondents. As 20% of the respondents have practically fallen out of the research (with their responses being only partly valid and thus discarded by the researchers), the authors intend to interview several representatives from both groups in order to reveal the cause of the problem so that it could be avoided in the future.

5   Research Results

The students' and the employees' responses were not unanimous, which reflects the respondents' voluntary and free expression of their opinions, without any prior preparation or instructions given but the request not to skip questions, express their true attitudes and not exceed the 20 min' time limit. The responses have provided a variety and inconsistency of the Russian students' knowledge of UAVs and their usage, though some of the results have proved to be quite opposite to the expected attitudes of university students and university employees.

5.1   Analysis of the Responses

The survey intended to reveal the primary attitudes, which could be analysed and generalized to serve as basis for further research and did not aim at identifying deep personal reasons of the students' and employees' reactions. Therefore, the questionnaire contained 16 questions – 3 background questions and 13 thematic multiple-choice questions, which yielded the following results considered below.

The background questions aimed at revealing the respondents' age, gender and occupation as this information has revealed their maturity and responsibility for these particular responses, partly explained their attitudes and gave ground for further research of the topic. It is evident from the histograms that most of the respondents are students – under 25 years of age with the standard deviation of about 5 years. The range of the age was found to be 38 years starting from 19 and up to 57 years of age. The data related to the respondents' age are presented in Fig. 1.

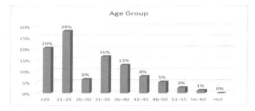

Fig. 1. Age groups

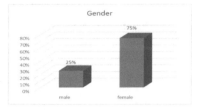

Fig. 2. Respondents' gender

Gender is an important variable in the given research situation which is variably affected by both the socio-economic situation in the country and the research situation in the given university in particular. Hence, though investigated for this study, gender was not further taken into account because of the highly predictable gender situation at the pedagogical university. The data related to gender issues are presented in Fig. 2.

The students' and employees' responses though diversified within each group have reflected more or less uniform attitudes in their responses to Questions 1–5 and Question 8; however, the research hypotheses required comparative analysis of the students' and employees' responses to some questions, which results are discussed in 5.2 below. As is seen in the histogram below, 72% of the respondents were students (57 persons), 14% were academics (11 persons) and 14% were university managers (11 persons). Although all the academics involved conduct research on a more or less

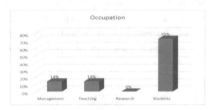

Fig. 3. Occupation

regular basis, their positions are not listed as researchers, so they did not indicate themselves as such, while researchers having no teaching workloads did not take part in the survey. In the absence of big enough separate samples of academics and managers, their responses were combined and analysed in one united sampling. The data are presented in Fig. 3.

Question 1 was intended to identify the general attitude of the respondents to the emergence of UAVs/drones. The responses revealed that 29% of the respondents approved and 8% fully approved the emergence of drones, 6% disapproved and 5% completely disapproved their emergence, with a majority of the respondents (52%) remaining neutral. The neutral majority indicates the respondents' absence of experience in this issue (they have never used or dealt with drones in any way) – see Fig. 4.

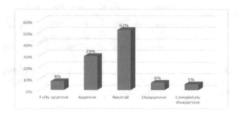

Fig. 4. General attitude to the emergence of drones

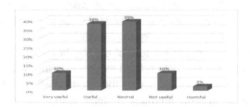

Fig. 5. Assessment of drones as useful/harmful

Question 2 aimed at revealing the respondents' attitude to the usefulness or harmfulness of UAVs/drones. The responses have revealed quite positive attitudes of most of the survey participants, with 10% finding drones very useful, 38% believing in their usefulness, 39% remaining neutral and only 10% considering drones useless, with 3% finding them harmful. It is worth noting that the respondents seeing drones as harmful were in the student group, and (quite unexpectedly) 12% of the students found drones useless against 4% of the employees, which reveals a higher level of professional systemic analysis of the academics and university managers (see Fig. 5).

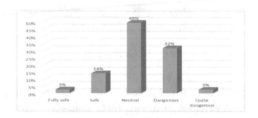

Fig. 6. Assessment of drones as safe/dangerous

Question 3 was supposed to identify the respondents' perception of the safety/danger category. The results have revealed a majority of neutral attitudes, which again must have proved the absence of the respondents' experience and their unwillingness to be involved in any issues concerning drones. The respondents considering drones fully safe (3%) proved to be students, with only 1% of the students finding drones quite dangerous against 4% of employees. 32% of all the respondents are confident about potential danger of drones, with only 14% thinking that they are safe (Fig. 6).

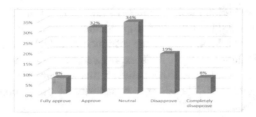

Fig. 7. Dis/Approval of drones applicability in commerce

Question 4 aimed at revealing the respondents' general attitude to the use of UAVs/drones in trade and commerce. Although 34% of the respondents expressed their neutral attitude (mainly due to the limited or no experience), 8% fully approved and 8% fully disapproved the use of drones in logistics. Some students gave full approval (with zero full approval by the employees which reveals their higher level of cautiousness and responsibility), but it was unexpected that 7% of the students fully disapproved the use of drones for commercial purposes (and only 4% of employees), with 15% of the students and 9% of the employees disapproving their use in trade (see Fig. 7).

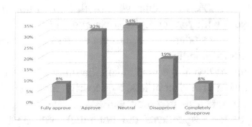

Fig. 8. Dis/Approval of drones use in armed conflicts

Question 5 was intended to find out the respondents' attitudes to the use of UAVs/drones in armed conflicts. The responses showed similar extremes – 8% fully approved and 8% completely disapproved the use of drones in armed conflicts, with 15% of the students and only 9% of the employees fully approving the use of drones in armed conflicts, and 12% of the students and 0% of the employees completely disapproving it. 28% of the students approved the application of drones in armed conflicts against 42% of the employees, which may reflect the employees' deeper emotional involvement in the recent and ongoing armed conflicts (Fig. 8).

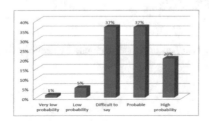

Fig. 9. Probability of drones abuse

The responses to Questions 6–7 and 9–13 are analysed in 5.2 below.

Question 8 was designed for the respondents' assessment of the probability of drones use by terrorists. With 37% of the respondents actually refusing to answer this question (option 'Difficult to say'), 37% expressed their concern for the probability of terrorist attacks with drones and 20% believed that the probability is very high, with only 6% feeling quite safe. Therefore, the majority of the respondents worry about possible abuse of drones (Fig. 9).

5.2   Comparative Results

The responses to Questions 6–7 and 9–13 obtained results, which were worth comparing as they revealed the ongoing political and legal discussions concerning the use of drones inside and outside hostilities and provided deeper understanding of the attitudes of two generations – academics and university managers in charge for the education of tomorrow's decision makers in Russia.

Question 6 was expected to clarify the respondents' attitude to the use of drones in military operations by national armed forces (the Armed Forces of the Russian Federation included). The employees proved to be more involved in the issue (18% neutral only against 33% of the students) and much more positive: 65% approving (against 40% of the students) and 13% fully approving (with 12% of the students), with 4% of disapproving employees with zero complete disapproval (8% of disapproving and 7% of completely disapproving students, which might express their disapproval of Russia's involvement in any armed conflicts).

Question 7 aspired to reveal the respondents' attitudes to the possibility of drones use in peacetime outside armed conflicts by security forces (FSS, CIA, FBI), both Russian and American taken as common examples. The results received were a bit unexpected – 34% of the respondents did not mind, with 32% approving security forces' use of drones and 8% fully approved it, with 19% of disapproval and 8% of complete disapproval. However, the comparative analysis has revealed a higher level of the Russian employees' approval – 51% (against 24% of approving students), with 12% of students completely disapproving against the employees' zero disapproval in this option (see Fig. 10).

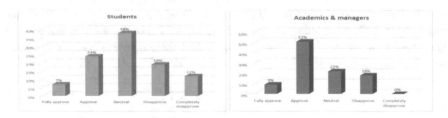

Fig. 10. Attitude to drones use by security forces

Question 9 meant to find the respondents' assessment of the probability of terrorist attacks using drones in peacetime outside armed conflicts. An overwhelming majority of the respondents found this probability real – 42% and even high – 24%, with 24% expressing their unwillingness to answer the question, and 10% of optimists considering the probability to be low (0% – very low). The employees showed deeper involvement – only 9% chose the 'Difficult to say' option against 29% of the students, who did not mostly assess the probability as low (5% against 22% of more optimistic employees – a very unexpected result), with a majority of students and employees assessing high the probability of terrorist attacks using drones in peacetime outside armed conflicts: probable – 38% of the students and 51% of the employees, highly probable – 28% and 18% accordingly.

Question 10 was intended to reveal the respondents' attitude to application of drones as both a new method and means of war. Although 38% of the respondents stated their attitude was neutral, the majority of the responses divided into two – 29% found the new method and means of war inhuman (with large civilian casualties) and less human (4%), while 13% considered them more human and 16% – quite useful (allowing to save soldiers). And here the students demonstrated deeper concern: 33% found the drone use inhuman against 18% of employees (less human – 1% and 9% accordingly), with 33% of the students being neutral against 51% of the employees, with 33% of the students and 22% of the employees expressing positive attitudes to the issue (more human – 14% against 9%, and quite useful – 19% and 13% accordingly).

Question 11 concerned the responsibility for the use of drones in hostilities and showed unanimity of the respondents in the two groups. 42% of the survey participants believed national governments were responsible (38% of the students and 51% of the employees), 30% put the responsibility on commanders-in-chief (31% and 27% accordingly), 13% found responsible commanders of operations (commanders on site) – 14% and 9% accordingly, and 14% considered drone operators fully responsible for the drone attacks (16% and 13% accordingly). The option 'Manufacturing corporations' was only chosen by 1% of the survey participants (1% of the students with 0% of the employees). These results have proved the need for more elaborated national and international legislation as seen by the surveyed Russian citizens.

Question 12 proposed to reveal the respondents' attitudes to the responsibility of drone operators for civilian casualties and yielded the following results: though 27% of the respondents chose the 'Neutral' option, an overwhelming majority believed that drone operators were responsible (34%) and fully responsible (30%) for civilian casualties, with 5% only believing in the low responsibility and 4% finding them not responsible. The compared results of the two groups turned out to be quite unexpected and very interesting for further research: an overwhelming majority of the respondents in both groups found drone operators responsible for civilian casualties: highly responsible – 31% of the students and 40% of the employees and fully responsible – 33% of the students and 29% of the employees, with 29% of the students and 18% of the employees preferring the 'Neutral' option. It was quite unexpected that 13% of the employees exempted drone operators of any responsibility (with 0% of the students), while 0% of the employees and 7% of the students chose the 'Low responsibility' option.

Question 13 aspired to highlight the respondents' understanding of the level at which decisions on the use of drones in military operations should be made. A majority (40%) considered the level of national governments to be the most appropriate one, 23% believed it was the level of international organisations (e.g. UNO), 12% chose the 'International Law' option, and 19% thought it was the level of commanders of military operations, with only 6% choosing military organisations (like NATO), which might have revealed the purely Russian understanding and attitude to this organisation, which has been constantly depicted as hostile to the Soviet and Russian national interests. The results of the two groups proved to be very close, with 42% of the students and 34% of the employees voting for national governments, 21% and 22% accordingly – for commanders of the military operations, 22% and 22% choosing international organi- sations, and 6% and 4% accordingly distrusting military organisations (e.g. NATO). As regards to the 'International Law' option, it was preferred by 18% of the employees and 9% of the students only, which most probably reflects the different level of education of the respondents in the two groups (in the absence of law students and teachers).

6   Conclusions and Discussion

The results of the legislation and discourse analysis supported by the survey findings have highlighted a scope of problems proving the correctness of the previously for- mulated economic, social, political, legal and organisational challenges facing the drone use and possible abuse of both UAVs and humans. Hypothesis 1 has been proved partly as the students participating in the survey have expressed a high enough level of cautiousness – very probably because they were mainly females. Hypothesis 2 has been proved, though the employees have also showed a high level of sensitivity to decisions taken by single persons (even experts) and a high enough level of distrust, which again might be explained by a majority of females in the employees' group.

The five challenges revealed above, facing the growing use of drones as both a prominent means of commerce and warfare and the results of the survey involving Russian students and university employees as stakeholders of Russia's policies future development, have prompted five priority issues highlighting the primary steps to be taken to resolve the existing problems in legal, political and organisational regulations of drone use in the present and future both in peacetime and armed conflicts.

Priority issue 1: The need for a comprehensive review of the current legal basis (both national and international) for drone application in commerce and warfare in order to pinpoint contradictions and inconsistencies and reveal possibilities for misuse and abuse of drones as means and methods of war and commerce (e.g. logistics).

Priority issue 2: The need for national and international changes to be introduced to the legal basis for the development of economic use of drones – at the level of WTO and other international organisations (ITC, ECO, WHO, AITIC, FAO, etc.) for the economic development and prosperity of the world to the fullest.

Priority issue 3: The need for deeper political involvement of all the drone stake- holders (80 countries) in order to prevent future clashes of economic and political interests and the use (and abuse) of drones as means and method of their resolution.

Priority issue 4: The need for the drone stakeholders/decision-makers uniting their efforts on international forums (like UNO, UNESCO) in order to reach a higher level of social stability and safety inside their societies and beyond them.

Priority issue 5: The need for International Law, IHRL and IHL reconsidered and amended to the necessary extent under the new conditions, to regulate the core terms of drone's production, sale and application at the global level and take the necessary organizational measures in order to reach global consensus as soon as possible.

References

1. Hetzenecker, J.: More than 3,000 Guests Attend Drone Racing League Event at BMW Welt. BMW Group.   https://www.press.bmwgroup.com/global/article/detail/T0283592EN/more-than-3-000-guests-attend-drone-racing-league-event-at-bmw-welt?language = en (2018)
2. Clarity from Above. PwC Global Report on the Commercial Application of Drone Technology. https://www.pwc.pl/en/publikacje/2016/clarity-from-above.html (2016)
3. Regulation of Drones. The Law Library of Congress, Global Legal Research Center. https://www.loc.gov/law/help/regulation-of-drones/index.php (2016). Accessed 21 Dec 2018
4. Qureshi, A.W.: The Legality and Conduct of Drone Attacks. Notre Dame J. Int. Comp. Law **7** (2), Article 5 (2017). https://scholarship.law.nd.edu/ndjicl/vol7/iss2/5
5. Karyakin, V.V.: UAVs – a new war reality. Prob. Natl. Strat. **3**(30) (2015). Available in Russian at https://riss.ru/images/pdf/journal/2015/3/10_.pdf
6. The Military Doctrine of the Russian Federation. Approved by the President of the Russian Federation on 25.12.2014 г. (# Pr-2976). http://base.garant.ru/70830556/ (2014)
7. Unmanned Systems Integrated Roadmap FY2017-2042. http://www.defensedaily.com/wp-content/uploads/post_attachment/206477.pdf (2017)
8. Branley, A., Armitage, R.: Perpetrators using drones to stalk victims in new age of technology fuelled harassment. https://www.abc.net.au/news/2018-10-01/drones-used-to-stalk-women-in-new-age-of-harassment/10297906 (2018)

Visual Detection of Trespassing of a Small Unmanned Aerial Vehicle into an Airspace

Kai Way Li[1(✉)], S. J. Chang[1], Lu Peng[2], and Caijun Zhao[3]

[1] Department of Industrial Management, Chung Hua University,
Hsin-Chu, Taiwan
kai@chu.edu.tw
[2] Department of System Engineering and Engineering Management,
City University of Hong Kong, Kowloon Tong, Hong Kong
[3] School of Safety and Environmental Engineering, Hunan Institute
of Technology, Hengyang, Hunan, People's Republic of China

Abstract. Small unmanned aerial vehicles (sUAV) have becoming widespread both in commercial and for entertainment usages. They have been used in both business and entertainment purposes. There are sUAV-induced problems. sUAV trespassing into prohibited areas has been the major one of them. A sUAV was guided using a remote controller to suspending in the air. The sUAV may enter a prohibited public area. It may also enter private areas unwelcomed. We conducted a study to quantify human subjects' sensitivity in discovering the trespassing of a sUAV into a specific area. Twenty human subjects were requested to determine whether the sUAV has trespassed in the test field or not by answering the question "whether the sUAV is inside the test field?" by replying a score from 1- definitely yes to 5- definitely no. The conditional probabilities of finding trespassing of the sUAV were calculated. A receivers' operating characteristics curve was plotted. The P(A) was calculated and was adopted to represent the sensitivity of the subjects in detecting the trespassing of the drone. The outcomes of the analysis of variance testing the effects of the direction on the probabilities of replying "definitely yes" and "definitely or probably yes" were both insignificant. Neither were the effects of direction on the P(A) significant.

Keywords: Unmanned aircraft vehicle · Drone ergonomics · Trespassing · Subjective rating · Signal detection

1 Introduction

In recent years, usage of small UAV (or sUAV) are becoming widespread for personal use. The attractiveness of the sUAV for the hobbyists and players has pushed up demand in the market sharply. In the PRC, the 2014 market sales of civil drones were as high as 1.5 billion RMB. This number was estimated to be as high as 60 billion in 2019 [1]. An internal record of the DJI, a company manufacturing quadcopters in China, showed that the company has sold over one million Phantom series quadcopters during 2013 to 2015. The aviation authority (FAA) of the USA began to implement a sUAV administration regulation on 2015 and obtained three hundred thousand

© Springer Nature Switzerland AG 2020
J. Chen (Ed.): AHFE 2019, AISC 962, pp. 230–237, 2020.
https://doi.org/10.1007/978-3-030-20467-9_21

enrollments in just one month [2]. The FAA [3] estimated the total UAV sales in the USA for consumers and hobbyists would increase from 2.5 million in 2016 to seven million in next four years. The sharp increase of sUAV has created public safety problems. Crashes of UAV were common [4, 5]. The crashing drone caused injuries and even death for persons on the land [6]. sUAV trespassing into the flight route of commercial jet near airport jeopardizing the jet taking off or prepared to landing have also been reported [7]. Such trespassing has resulted in losses not only to the airline carriers but also to our society [8]. UAVs trespassing for aero photography also raised the public concerns of privacy. A sUAV flying to a private property talking video or photos is definitely unwelcomed. Government Regulations for sUAV operation have been announced because of public safety concerns.

For public safety and security reasons, the aviation administration authorities have declared non-fly zone and restricted-fly area for sUAV [9, 10]. sUAV flying at altitude higher than 60 m are prohibited in any restricted fly area [6]. The sUAV navigators need to identify those zones mostly via visual monitoring of the sUAV and mapping to ground areas, so as not to violate the regulations. Trespassing of a sUAV into an field is common [2, 7]. A sUAV flying into the borderline of a property on the land lead to a trespassing. The majority of the sUAV are so small. Sophisticate equipment may not detect a sUAV trespassing because it is too small. However, the capability of human in detecting the trespassing of a sUAV in the field has not been explored. A study to investigate the sensitivity of human subjects in finding sUAV trespassing in the field is significant.

The signal detection theory has been adopted in analyzing the sensitivity of human detectors in detecting presence of signal [11]. In the theory, the responses of a detector in detecting signal include hit, miss, false alarm, and correction. In sUAV detection, a hit occurs when the sUAV cross the borderline and fly into a field and the detector respond a trespass. A miss happens if the detector answers no trespassing. When a sUAV is outside the field and the detector answers yes the sUAV is trespassing, a false alarm appears. The answer of no trespassing, when a sUAV is outside the filed, leads to a correct rejection. We performed this study to determine the capability of ground observers in discovering the trespassing of a sUAV into a specified area.

2   Methods

2.1   Human Subjects

We recruited twenty adults as human subjects in the study. Fourteen of them were male and six were female. They were 22.1 ($\pm$1.6) years old. A visual Acuity measuring device (Stereo Optical, Chicago) was employed to measure the subjects' visual acuity. Our subjects had near normal vision and normal color discrimination capability. Their averaged corrected visual acuity for both right and left eye were 0.90. The subjects had normal auditory capability. We obtained signatures of the informed consent for all the subjects before they attended the experiment.

2.2   Unmanned Aerial Vehicle

A Phantom 4 sUAV (DJI®, China) was adopted. The navigator controls the sUAV remotely using a controller. A cellphone (iPhone® 6 Plus) was used as the displayer. The sUAV navigator may read the position of the sUAV in the air from the displayer. The distance of the sUAV to the taking off position and the sUAV altitude when flying are displayed in digits. This Phantom 4 may station in a specific position in the air. This occurs when the navigator release both joys tickers of the remote controller. In other words, the Phantom 4 will stay in the same position in the air when there is no fly command. When the drone moves away from the position because of the wind, it flies back and stays in the same position. This function depends on the positioning system. When strong GPS signal is received, this signal is used in the positioning of the vehicle in the air.

2.3   Test-Field

The test field was in the sport field (see Fig. 1). In the sport field, the arc of the out rim of the runway was selected as the borderline of the field. There were three directions. On each direction, there were 16 positions (+ in Fig. 1). These positions were composed of two distances of the sUAV to the borderline either within or outside the test field and four sUAV altitudes. Figure 2 shows the positions of the sUAV (+) in direction B. sUAV positions were also planned in the other two directions. There were, therefore, forty eight sUAV positions in total.

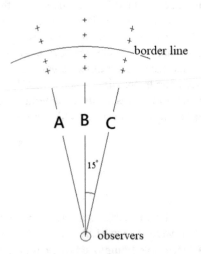

Fig. 1. Test-site in a university stadium

We conducted the experiment in the morning. The subjects might watch the test field without glare. The visibility was 10.0 km. The sky was clear and there was sunshine and light breeze. The background noise was 47.0 dB. A light meter (Lux-Meter, TransInstrument, Singapore) was used to measure the illuminance at the origin

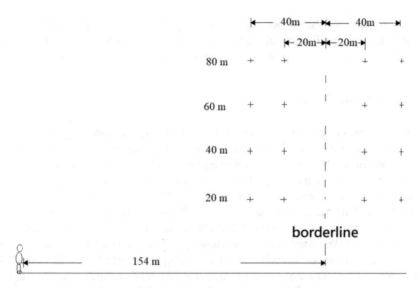

Fig. 2. Altitudes of drone hovering positions

on the ground. The reading was 1050.0 lx. The humidity and temperature were 75.0% and 30.0 °C, respectively. The wind speed, temperature, humidity, and visibility were reported by an app UAV forecast on the smartphone.

2.4 Procedure

Before the experiment, the subjects were in the origin. The research personnel told then that they would watch a sUAV in the front and to see if the sUAV has trespassed the test field. The edge of the runway was the borderline of the test field. The sUAV has trespassed if it was inside the borderline. Otherwise, a trespassing has not happened. They could not see when the navigator was navigating the sUAV to one of the positions in the air. When the sUAV has arrived at a predetermined position and was suspending in the air, we informed the subjects to watch the sUAV. They needed to determine whether the sUAV has trespassed or not and answer a score from 1- definitely yes to 5- definitely no. The subjects could not talk to others. Neither could they watch others' answer. After all of the subjects had done, the navigator guided the sUAV to the next position without the witness of the subjects.

2.5 Data Analyses

There were nine hundred and sixty rating scale tasks for the sUAV detections. The answers of the subjects were summarized. The conditional probabilities of finding the target when the sUAV was trespassing and when it was not within the test field were calculated. We plotted the receiver's operating characteristic curves under horizontal distance and altitude conditions. The sensitivity of the subjects, or P(A), were analyzed using the receiver's operating characteristic curves [11]. In addition to the sensitivity, the bias of the subjects in detecting trespassing of the sUAV was calculated. We adopted the least significance difference (LSD) test for posterior comparisons. We used the SAS® 9.4 software to do analyses for descriptive statistics and variance.

3 Results and Discussion

Figure 3 shows the hit rate or alternatively probability of answering yes when the sUAV was inside the field. When the responses of yes-definitely and yes-probability were combined, the hit rate was all 0.8 or higher for all distance and altitude conditions except at the 20 m distance and altitude of 60 m (0.68). The 40 m altitude and 40 m distance condition had the highest hit rate (0.92). At the altitude of 80 m, the hit rates were 0.8 for both the distance of 20 m and 40 m. When we considered the "yes-definitely" responses only, the probabilities of answering yes were relatively lower than those of the yes-definitely and yes-probability combined conditions. For the "yes-definitely" conditions, the probabilities of answering yes ranged from 0.22 to 0.67. The probability 0.22 occurred at the 20 m distance and 60 m altitude condition and the probability 0.67 occurred at the 40 m distance and 40 m altitude condition. The standard deviations for the yes-definitely and yes-definitely combined with "yes-probable" conditions were in the ranges of 0.03 to 0.23 and 0.03 to 0.26, respectively.

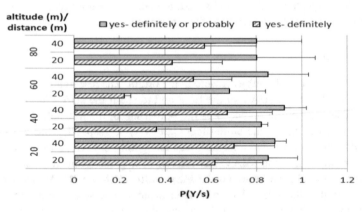

Fig. 3. Hit rate

Whether the probabilities of answering yes in directions of A, B, and C were compared. We examined both the probabilities of answering yes for "yes-definitely" and "yes-definitely" combined with "yes-probability" conditions. All these tests showed insignificant results. The effects of distance to borderline were tested. This variable was found significant ($p < 0.05$). We also examined the effects of sUAV altitude on the probability of answer yes. This variable was, however, insignificant.

Figure 4 shows the false alarm rate, or the probabilities of answering yes when the sUAV was not within the field. This probability was in the range of 0.02 to 0.13 when considered the yes-definite responses only. The probability 0.13 occurred at the 20 m distance and 40 m altitude condition. The range of the false alarm rate was between 0.12 to 0.42 when both yes-definitely and yes-probably conditions were considered. The highest false alarm rate occurred at the 20 m distance and 40 m altitude condition. It was not clear why the subjects had the highest false alarm rate when the sUAV was 40 m above ground level and was 20 m to the borderline. The ranges of the standard deviations of the false alarm rate considering the yes-definitely and the yes-definitely-or-probably conditions were between 0 and 0.13 and between 0.06 and 0.2, respectively.

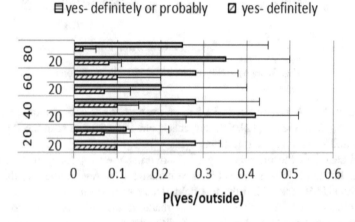

Fig. 4. False alarm rate

The probabilities of answering no when the sUAV was not entering the field, or correct rejection rate, were calculated. When considered only the yes-definitely condition, the probabilities was in the range of 0.13 to 0.56. When both the no-definitely and no-probably were considered, the range was between 0.53 and 0.78. It was apparent that the subjects had much higher correct rejection rates when they were allowed to respond no including both the definitely and probably decisions.

We plotted a receiver's operating characteristic curve for each of our experimental condition. The P(A) were determined. Figure 5 shows the averaged P(A) under experimental conditions. At the distance of 20 m, the P(A) for the altitudes of 20 m, 40 m, 60 m, and 80 m were 0.857, 0.735, 0.691, and 0.772, respectively. When the sUAV was 40 m away from the edge of the runway, the P(A) for these altitudes were 0.90, 0.869, 0.825, and 0.842, respectively. The P(A) when the distance was 20 m to the borderline (0.76) was significantly ($p < 0.05$) lower than that when the sUAV was 40 m (0.86) away. The stand deviations of the P(A) for the 20 m distance condition was between 0.06 to 0.11. For the 40 m condition, they were between 0.04 to 0.16.

Detecting the trespassing of a sUAV is important for safety and security for the public. In our experiment, the subjects could not see before the sUAV had reached a certain position. They watched to determine if a trespassing had happened or not. They needed to have quick eye movements and fixations of the eyes when scanning the sky

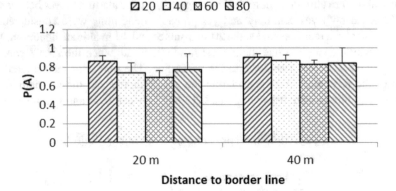

Fig. 5. P(A) under distance and altitude conditions

to find the sUAV. A person needs 12.5 s, on average, to search, decide, and to act to escape an impact in the air [12]. Our subjects spent 10 to 20 s to scan the sky to search the drone and to answer the question.

Visual search is the primary way of discovering the trespassing of a sUAV by a person on the land. When the subjects watched the sUAV in the air, they judged whether the sUAV was within the test field. There were no markers or other physical objects helping them to judgement. This made it difficult for them to make the judgement. We believed that they had more difficult when the drone was away from the borderline further and when the drone was at a higher altitude. They needed to utilize their sense of spatial perception of the runway and the drone. Sound of the drone might provide cue for trespassing judgement.

Conducting a study of sUAV trespassing detection in the field is extremely complicated because of the difficulties in controlling the environmental parameters and conditions. In our study, the experiment was conducted in the field in a stadium in the morning. The weather was fine and the sky was clear with light breeze. We believe that our test conditions represent the most favorable weather and visibility conditions for sUAV trespassing detection. Our results might be totally different if the visibility is low. Even though we know the suspending accuracies of the sUAV from the product information of the Phantom 4, the suspending accuracies of the Phantom 4 in the air are surely rely on the speed and direction of the wind. They also rely on satellite signal. The speed of wind might change abruptly during the experiment. The errors of the reading of sUAV position in the air during testing were, actually, unknown. In addition to the difficult of controlling the environmental parameters, gathering the human subjects to the field was also not easy. There might also be interruptions due to people doing exercise in the sport field. The research personnel had to prevent the interruption from irrelevant people.

Acknowledgements. This research was financially supported by a grant from the Ministry of Science & Technology (MOST) of the Republic of China (under contract MOST106-2221-E-216-008-MY3).

References

1. Global UAV net. Trend and predictions of UAV market in China. Retrieved from http://www.81uav.cn/uav-news/201609/06/19009.html/ (in Chinese)
2. Luppicini, R., So, A.: A technoethical review of commercial drone use in the context of governance, ethics, and privacy. Technol. Soc. **46**, 109–119 (2016)
3. Federal Aviation Administration. FAA aerospace forecast: Fiscal years 2016–2036. Retrieved from http://www.faa.gov/news/updates/?newsId=85227&cid=TW414/
4. Williams, K.W.: A summary of unmanned aircraft accident/incident data: human factors implications. Technical report DOT/FAA/AM-04/24, Federal Aviation Administration, US Department of Transportation
5. Democrat & Chronicle. Domestic drone accidents. Democrat & Chronicle. Retrieved from http://rochester.nydatabases.com/map/domestic-drone-accidents/
6. Forest C.12 drone disasters that show why the FAA hates drones, TechRepublic webpage. Retrieved from https://www.techrepublic.com/article/12-drone-disasters-that-show-why-the-faa-hates-drones/
7. Rao, B., Gopi, G.A., Maione, R.: The societal impact of commercial drones. Technol. Soc. **45**, 83–90 (2016)
8. Loffi, J.M., Wallace, R.J., Jacob, J.D., Dunlap, J.C.: Seeing the threat: pilot visual detection of small unmanned aircraft systems in visual meteorological conditions. Int. J. Aviat., Aeronaut., Aerosp. **3**(3), 1–26 (2016)
9. Civil Aeronautics Administration. No-fly and restricted fly zones for unmanned aircraft vehicles near airports (2017). Retrieved from https://www.caa.gov.tw/en/search/searchResult.asp?cx=000286267817699135961%3Akpxubvxz79m&ie=UTF-8&q=UAV&x=0&y=0&as_fid=4aOZ4neHn5QXzKmNf55/
10. Li, X.C.: Summaries of 10 dangerous UAV flying cases. Retrieved from http://www.gg-robot.com/asdisp2-65b095fb-59748-.html (in Chinese)
11. McNicol, D.: A primer of signal detection theory. Allen & Unwin, London (1972)
12. Federal Aviation Administration. Pilots' role in collision avoidance. Retrieved from http://www.faa.gov/documentLibrary/media/Advisory_Circular/AC_90-48D.pdf

Issues of Human Error Management in Introducing New Technology for Bridge Inspection Drones

Riku Tsunori[✉] and Yusaku Okada

Department of Administration Engineering, Faculty of Science and Technology,
Keio University, Kanagawa, Japan
rikutsunori@keio.jp, okada@ae.keio.ac.jp

Abstract. When introducing new technology, operators need to reduce the risk to an acceptable level for society while careful not to lose the merit of new technology to create innovation. The site tends to rely on fail-safe and foolproof, it is possible to deal with the event that happened once, but it cannot respond to unexpected events. Therefore, it is important to conduct preventive measures to improve safety even on human resources base.

Keywords: Unmanned aerial vehicles · Safety management · Risk

1 Introduction

In recent years, unmanned aerial vehicles are spreading rapidly, and their use has come to various fields such as inspection, security, logistics, etc. in addition to conventional aerial photography and agricultural chemical scattering. It can realize labor-saving, unattended, low cost, etc. It is because it leads to improvement and resolution of problems that had been involved, such as chronic labor shortage and soaring labor costs. This change, also called the empty industrial revolution, is very pleasing as it is considered to enhance the convenience and affluence of our lives. Meanwhile, with the introduction of unmanned aerial vehicles, safety problems such as dropping, or collision of the airframe are occurring. Third party damage such as workers' accident that the aircraft becomes uncontrollable and causes cuts on the parties involved, the airframe lost during flight, falling nearby households, and dam-aging the roof has also occurred. In order to use unmanned aerial vehicles without causing troubles exceeding the allowable amount to society's risk, safety management in the field operation is important. However, in the field, priority is given to the progress of work, safety is often secondary. Also, prior to adequate discussion on safety, new technologies have been constructed and technologies are sometimes deployed without safety improvement. As a background to responding to such safety, there is a way of thinking about safety based on the prevention of reoccurrence. Regarding troubles that have never occurred, there is no information to think about countermeasures, and I do not know whether trouble will occur in the future. Since there is no guarantee that effects will be obtained even if you spend time or money such as prevention beforehand, it is not guaranteed that it will be effective, so we only consider prevention of recurrence, and

© Springer Nature Switzerland AG 2020
J. Chen (Ed.): AHFE 2019, AISC 962, pp. 238–245, 2020.
https://doi.org/10.1007/978-3-030-20467-9_22

regarding the troubles that have not yet happened, often it is. In Japan before, there was atmosphere that was tolerable that it cannot be helped if it cannot cope with unexpected troubles. However, the tolerance to safety is high now, and such a sweet idea is not allowed. In order not to lower the motivation to introduce technology, it is necessary to prevent occurrence of troubles, at least to minimize the damage. However, if you do excessive safety measures, if workability gets worse or the goodness of new technology is lost, there is no motivation to introduce in the first place. Therefore, in this research, from the viewpoint of operation management at the site, we decided to organize the safety issues related to drone use and to consider safety management that was balanced with the merits of the new technology.

2 Analysis on Safety Management of Drone Use in General

Analysis was carried out based on the data released by the Ministry of Land, Infrastructure, Transport and Tourism and the Ministry of Agriculture, Forestry and Fisheries in order to arrange the current problems concerning troubles in using drone and its countermeasures.

2.1 Analysis of Characteristics of Accidents and Troubles (Ministry of Land, Infrastructure and Transport Data from 2016 to 2019)

As a person who caused trouble, individuals and aerial photographers are ranked high, and these two occupy less than 60% of troubles that occurred. First of all, looking at individuals, it is certain that people with little handling experience are causing troubles many times, but people who have more than 100 h of maneuvering time have also caused trouble. Therefore, it is impossible to say that individuals are in trouble because of less piloting experiences. In order to grasp the difference in cause of trouble due to the difference in the maneuvering time, the cause of the trouble was largely divided into five. Although I thought to analyze from the viewpoint of 4M used for analysis of accident factors, there were many data for which the pursuit of the factor was insufficient and there were some that only wrote the contents of the trouble. Therefore, the cause of the problem (Cause 1), the skill/knowledge of the worker, the problem related to improper work (Cause 2), the problem of the flight environment except cause concerning radio waves and magnetism (Cause 3), the problem of mechanical trouble (Cause 4), problems related to radio waves and magnetism (Cause 5). Also, with reference to the average maneuvering time, we decided to divide into two groups by maneuvering time and compare each. As a result, it was found that there is a difference regarding the ratio of Cause 2 to Cause 4 when the maneuvering time is over 60 h and less than 60 h. If you are using unmanned aerial vehicles and you are receiving sufficient education and training as done in a large enterprise, in case of using by an individual, such things are hard to imagine. Therefore, it seems that there are many times that you actually use it while the skill and knowledge of the pilot and assistants are insufficient. It is difficult from a variety of viewpoints such as personnel, cost, accumulated expertise, information gathering ability, and the like to seek to undertake efforts at the same level as large companies. However, if you work safely in the same way as before, there may be a serious accident such as giving a serious injury or putting out the dead, which will encourage the uneasiness of the public. It is important to consider incentives to become more enthusiastic about safety, administration such as national and local governments and associations concerning drone show incentives and encourage initiatives for safety It will be necessary. Next, looking at the aerial photographing related businesses, people with over 100 h who are considered to have a lot of handling experiences have caused troubles a lot, and cause 2 for troubles is low because the operation time is long It never came. It seems that the characteristics of aerial photography are involved in this. It is an advantage of using a drone to be able to easily shoot pictures that required difficult photography from the ground or that required a costly helicopter or ship. Also, since there is little worry that the photographer is threatened with danger, it is also used in hazardous areas like volcano observation. However, due to this ease, it is considered that taking photographs taking risks lightly and neglecting safety. Often the place of use is not a densely populated area, and it is one factor that leads to safety disrespectation because it does not have to consider much the harm to others. In addition to individual and aerial photographers, we also analyzed the cause of the trouble, including construction related companies, local governments, agricultural relations and research institutes. As the whole feature, Cause 2 is increasing, so when thinking about trouble factors, it can be said that it is difficult for the eyes to go besides the pilot who caused the trouble (· assistant). Next, the person who made the flight was divided into four types of personal type, risk type, profit type, administrative/non-profit type, and analyzed according to its characteristics. Then you can see that there are much Cause 2 of type b, and there are much Cause 4 of type c. First, although it is type b, it is a type that takes risks in order to take pictures that you want. By taking risks, we needed more maneuvering skills and thought that the troubles caused by the pilot were more likely to occur. Next, although it is type c, they all serve customers, and occurrence of trouble leads to a decrease in reliability. In order to maintain business relations without losing trust when trouble occurs, it is important to show the cause of the trouble and to show the attitude to take proper measures properly. Cause 1 of type c is less than others, and maybe it reflects the attitude that we are sure to clarify the cause. The reason why Cause 4 has increased is that most companies are companies with strengths in technology, and it is considered that they are excellent in investigating the cause of aircraft trouble. To classify measures, I thought at 5E. However, since we decided to consider measures for the dragon's flight environment not as Environment but as Engineering classification, we decided to classify it from 5E as 4E excluding Environment. As a whole feature, although the percentage of Enforcement increased, it is thought that it was done as an improvement method because safety management was insufficient and operation or work which should originally be done was not performed. Next, looking at each type, it is characteristic that the ratio of Engineering of type c is large. This duplicates the analysis of the cause of trouble, but for type c which provides the service to the customer, it may lose the trust depending on the response at the occurrence of the trouble. Basically, the countermeasure of Engineering is a measure of fail-safe, foolproof, and it is a measure to prevent recurrence in a visible form, so I thought that it is easy to convey to customers that we are working firmly on safety.

2.2   Formulas Analysis of Characteristics of Accidents and Troubles (Ministry of Agriculture, Forestry and Fisheries Data from 2015 to 2017)

Speaking of helicopter overseas, mostly military use is done, but in Japan there is a history that it has been used for agricultural chemicals for many years. As of the end of February, 2018, the number of registered aircraft is said to be about 2,800, and the number of authorized operators will be about 11 thousand. Paying attention to the safety efforts of unmanned aerial vehicles in the field of agriculture where accumulation of know-how exists, I thought that it would be worth learning from current problems and countermeasures against it. Therefore, I thought about analyzing based on accident information of unmanned aerial aircraft and report by Ministry of Agriculture and Forestry concerning safety measures. There were 49 cases of contact accidents to overhead lines etc. in 53 cases out of 2015, 52 cases out of 62 cases in 2016, 57 cases out of 65 cases in 2017, accounting for the majority of accident of pesticide spraying. The thing to think about in terms of collisions with obstacles is whether you are making a flight plan after making thorough research in advance. Among the information that was being disclosed, we classified the cause of the trouble, but "overlooking obstacles due to lack of prior confirmation", "lack of cooperation between operator and navigator", "operator mistakes in operation, mistimetry", "Inadequate altitude, method, etc.", "Other (slip legs, gusts, breakdown of communication equipment, etc.)" were used. The first shortage of confirmation in the first cause is most related to the presence or absence of preliminary survey. The purpose of the preliminary survey is to clarify various risks such as existence of an obstacle, the condition of the airport, the property that may interfere with communications and radio waves, which is not understood by desktop assumption alone. In addition, each aims to deepen the understanding of the flight plan and have a certain common awareness. As long as we can have a common understanding of how to deal with the assumed risk, it will be difficult for the conductor to wait for instructions and each will be able to think and act instantaneously. Second, there is lack of cooperation between operator and navigator. As background of lack of cooperation, it is said that the sharing was insufficient about management problems such as the fact that the pilot judged without permission without knowing the importance of cooperation or without the assistance of the assistant, and how to cooperate Two problems are conceivable. The latter problem has much to do with the preliminary survey. Although it can be done on a desk if it is only to discuss general notes and roles related to collaboration, the importance of the preliminary survey arises when considering a cooperation method suitable for the situations in the field. It is necessary to think about correspondence according to the information obtained by going to the site. Cause There is a part due to lack of cooperation where the third measurement error and the altitude of the fourth flight are inappropriate. The question of slipping legs included in the other five things also may have been able to be noticed beforehand by understanding the badness of the footing in advance by walking the flight path in advance investigation. It is very helpful for us to think about what kind of accident is occurring and what is causing it as a reason to think about what to watch out for practical work, as well-known, it will be effective. However, in order to improve the site, it is important to further analyze the cause and pursue factors. If you do not do it, it will be such a story that you work hard, cooperate, be careful not to make a mistake, effort, be careful.

3 Issues of Safety Management on Site

Various efforts are underway to utilize drone in Japan, one of which is the Strategic Innovation Creation Program (hereinafter referred to as SIP) by the Cabinet Office. One of the multiple SIP programs aimed at realizing the safety and security of infrastructure with technology and conducted a demonstration experiment using drone for bridge inspection in various parts of the country. By going to the field of demonstration experiments and observing the work by various companies, we observe the current situation of safety management and the differences in approaches among companies, and based on the tasks obtained from that, We decided to analyze and propose management issues.

3.1 Risk of Using Drone

There are two risks concerning drone use. One thing is the risk of skipping drones. It refers to the influence of trouble that can occur in the process of flight preparation, takeoff, flight, landing etc. of the aircraft, such as workers' injury, third party damage, the influence on the natural environment, monetary influence due to aircraft damage. The second is the risk of stability as work. This is an instability factor given to work by using a drone, which is, for example, rainy weather, forced to interrupt the flight, and the business will not proceed as planned. Here we consider the former among the two risks. As mentioned earlier, the risk of the former is diverse, but the important thing to do on site work is the idea of human life first, not to harm workers and third parties. Other risks will be second, third as priority.

3.2 Response to Workers' Injuries

Examples of troubles of workers' accidents include electric shock by touching cables, electric shocks to lift a heavy aircraft, falling from the bridge, falling off the bridge, drone that is automatically returning collided with the crane and injured the workers at the falling place It can be cited. Every time such a trouble occurs trouble analysis, measures to prevent recurrence such as fail-safe and fool proof are being carried out, but the analysis is shallow, there is a problem that it cannot pursue up to the PSF level. In this background, there is a problem of relying too much on fail-safe countermeasures and a problem of not turning to regular operation when trouble occurs. The former will be described later, so we will first describe the problem that does not turn to regular work. PSF is a behavior formation factor and it is a factor that affects the value of human behavior such as human reliability, work efficiency, workload, or burden. In the explanation of A. D. Swain, it is the situation characteristic, the form of instruction of job and work, the characteristics of work and equipment, psychological stressor, physiological stressor, personal factors and so on. In thinking about these behavior formation factors, it is important to pay attention to the routine tasks of stationary tasks. Of course, it is necessary to think about unsteady tasks in which trouble has occurred, but if we pay attention only to that, we can only think about factors that exist only in nonstationary work. To make countermeasures every time trouble is to make several barrier walls against danger. In order to enhance safety, it is necessary to cover the leakage of protective walls, but in doing so, we will conduct risk assessment considering the error handling of people involved and the possibility of error, and for the assumed human error It is important to respond in various forms. Fail-safe measures are only for dealing with troubles that have occurred once, and it is human to respond when an unexpected event occurs. Fail-safe tends to depend on usual, and if the worker seems to be able to do business only in accordance with the manual, when an event that cannot be dealt with manually occurs, it will be in an instruction wait state and will not be able to move until something is told.

3.3   Response to Third-Party Damage

Examples of troubles caused by third party damage may be those in which the airframe becomes uncontrollable and collides with a third party, or equipment is blown by the wind and secondary damage is caused by the wind, the correspondence which is done is limited to the influence by the aircraft trouble. Third parties to be bothered by bridge inspection work are cars running on roads on bridges, sidewalks and passers-by riverbeds. By preventing the loss by the lead wire and preventing the drone from jumping out from under the bridge, it is possible to cope with troubles caused by aircraft to passengers and pedestrians. In addition to this, thorough monitoring of the surroundings and limiting invasion by third parties can also be used to countermeasure passersby under the bridge. Meanwhile, there are problems with handling other than problems caused by the aircraft. Workers' sensitivity to danger is low, and intuitive risk assessment at the work stage is inadequate. I have not been able to feel uncomfortable or uneasy about the dangerous situation that should be noticed if it is original. It is necessary to be able to imagine what kind of harm may be caused by a certain behavior or condition.

3.4   Basic Discipline to Workers

There is a problem that the discipline which is commonplace is not carried out at the work site such as railroad and general contractor where safety management was thoroughly in place where measures based on human resources are not made. Proceed in a straightforward manner, loudly pointing out and calling for confirmation, not being able to tell you thank you when you call me a voice call. The background of this problem is that we do not know the way of education even if noticed by the managers at the site, even if they are not aware of the fact that they do not have a natural discipline, and they have not been able to penetrate the site Sometimes. The fact that we do not have a normal discipline means that our consciousness to safety is inadequate. It is impossible to understand the effects such as organizing order, pointing to a finger, or communicating appreciation, but once you are taught its importance in training or work, you will not be able to master it. If you were educating "do it as a manual", you would not understand the reverse side of things that are basic methods of practice or being taken as measures against safety. There should be circumstances in each manual and notes on manuals, and if understanding can be educated along with the background, it may be easier to understand. For that purpose, it is necessary to create a manual and leave the background as a record of corrections in the middle. Moreover, even though I think that the awareness to safety is insufficient, the person himself may answer "I am doing business with safety aware". In such a case, it is good to listen to the background, taking as an example the safety measures that are done somewhat like normal. For example, asking why you are wearing protective equipment, most workers will answer that they are dangerous. One of the purposes for wearing protective equipment is mode change. I am going to the dangerous area from now on so that I can recognize myself about doing work. Also, although this will be from a different perspective, there is also the idea of wearing it for everyone. With the idea of wearing protective equipment because it is dangerous, if you think that you do not mind even if you injured himself, you do not have to wear it. In such a case, you should tell the effect of one worker being absent from work due to occupational injury. Knowing how much trouble the other people will incur if they cause workers' injuries will be a consciousness that they will be more safely and carefully taken from now on.

Not knowing the back side of safety measures applies not only to workers but also to administrators. Since the administrator does not understand properly, it is guided to subordinates "because it is a rule", "because it is defined by manual". I think that it is necessary to review from the part of whether the manager's safety awareness is sufficient to carry out the ordinary discipline.

3.5 Gathering and Sharing Information Within the Same Industry

Each company thinks that disclosure of trouble information will have a negative influence on the spread of new technology, basically setting trouble information private. Therefore, similar troubles frequently occur in inspection work of each business operator. Also, there are few places where information is gathered for important collections as a preliminary stage of information sharing. In addition to the safety of workers and third parties, the information on incidentally contains information that affects the efficiency and effectiveness of work, as well as information that affects the satisfaction of employees, and can be various hints for considering risk management. Therefore, collection of incorrect hits should be done. However, what you have to pay attention to here is to avoid making activities that just increase the burden on the site. With the ambiguity of what kind of effect will be brought to the site by collecting incorrect hits, the site will not move on its own initiative and a divergence from the manager will be born. Even submitting incident information, if you think that the situation will not change, you will lose motivation to submit. It is important for the industry as a whole, to build safety with few failure experiences because the world does not have uneasiness about drone. Through cross-ties within the industry, it is necessary to share trouble information, cooperate in factor analysis, measures planning, etc. so that development and safety are organically connected.

4   Conclusion and Future Prospects

The important thing in safety management in the introduction of new technology is to reduce the risk to a level acceptable to society by taking measures against safety and at the same time, not to lose the umami of new technology to become innovation. Currently, bridge inspection work by drone is still in the stage of developing and improving technology aiming at a level recognized by the country, and consideration for safety is inadequate. In the future, in addition to measures to prevent recurrence such as fail-safe, we will incorporate the concept of prevention measures, and to promote information sharing within the industry as to support human resources base and on-site safety management If we are able to do our own safety initiatives, drone usage will be tolerated as a way to make people's lives better. In this research, I have qualitatively discussed issues related to human error management in the introduction of new technologies based on observing the site of bridge inspection using drone. Because the history of using drone is still shallow, trouble data on bridge inspection using drone was not collected enough and it was because it was not open to the public. I am still considering how to do it, and I would like to bring it into a quantitative story by experiments and simulations.

References

1. Yusaku, O.: Human Factors Gairon. Keio University Press Inc, 225 (2005)
2. Ministry of Land, Infrastructure, Transport and Tourism. http://www.mlit.go.jp/koku/koku_tk10_000003.html
3. Ministry of Agriculture, Forestry and Fisheries. http://www.maff.go.jp/j/syouan/syokubo/gaicyu/g_kouku_zigyo/mizin_tuuti/index.html

Author Index

© Springer Nature Switzerland AG 2020
J. Chen (Ed.): AHFE 2019, AISC 962, pp. 247–248, 2020.
https://doi.org/10.1007/978-3-030-20467-9

Printed in the United States
By Bookmasters